32 ELECTRONIC POWER SUPPLY PROJECTS

BY ROBERT J. TRAISTER

TAB BOOKS Inc.
BLUE RIDGE SUMMIT, PA. 17214

FIRST EDITION

FIRST PRINTING

Copyright © 1982 by TAB BOOKS Inc.

Printed in the United States of America

Reproduction or publication of the content in any manner, without express permission of the publisher, is prohibited. No liability is assumed with respect to the use of the information herein.

Library of Congress Cataloging in Publication Data

Traister, Robert J.
   32 electronic power supply projects.

   Includes index.
   1. Electronic apparatus and appliances—Power supply. I. Title.
TK7868.P6T7   1982   621.381′044   82-5922
ISBN 0-8306-2486-4   AACR2
ISBN 0-8306-1486-9 (pbk.)

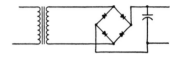

Contents

List of Projects   v

Introduction   vii

1   Dc Power Supplies   1
Voltage—Current—Summary

2   The Basic Dc Power Supply   8
Rectification—Diodes—PN Junction Diode Rectifier Circuits—Half-Wave Rectifier—Full-Wave Rectifier—Bridge Rectifier—Junction Diode Considerations—Power Transformers—Power Supply Filters—Bleeder Resistor Voltage Divider—Full-Wave Bridge—Full-Wave Combinational—Voltage Multipliers—Half-Wave Voltage Doubler—Full-Wave Voltage Doubler—Voltage Tripler—Voltage Quadrupler—Summary

3   Dc Power Supply Components   39
Transformers—Transformer Power Ratings—Tapped Primary Windings—Combining Transformers—Series And Parallel Connection of Transformers—Low Voltage Operation—Solid-State Rectifiers—Protective Components—Capacitance—Types of Capacitors—Color Codes for Capacitors—Capacitors for Power Supplies—Capacitors in Parallel—Capacitors in Series—Capacitor Insulation—Bleeder Resistors—Semiconductor Devices—Integrated Circuits—Light-Sensitive Solid-State Devices—Solar Cells—Summary

4 Voltage Regulators   103
Shunt Regulator—Zener Voltage Regulator—Solid-State Shunt Voltage Regulator—Electron Tube Shunt Voltage Regulator—Series Regulator—Solid-State Series Voltage Regulator—Electron Tube Series Voltage Regulator—Shunt Detected Series Voltage Regulator—Solid-State Shunt-Detected Series Voltage Regulator—Electron Tube Shunt-Detected Series Voltage Regulator—Current Regulator

5 Obtaining and Referencing Components   118
Cross-Referencing—The Experimenter's Junk Box—Keeping Track of Electronic Components—Summary

6 32 Electronic Power Supply Projects   133

Appendix A Schematic Symbols   285

Appendix B Wire Size and Current Rating   287

Appendix C Resistor Color Code   288

Index   289

LIST OF PROJECTS

Project 1: Half-Wave Power Supply	136
Project 2: Full-Wave Center-Tapped Supply	142
Project 3: Full-Wave Bridge Supply	146
Project 4: Add-On Zener Diode Regulator	149
Project 5: Dual Voltage Power Supply	153
Project 6: Dual-Polarity Regulated 15-Volt Supply	158
Project 7: 9-Volt Series-Regulated Supply	163
Project 8: Power Supply/Battery Charger	170
Project 9: Versatile Voltage Doubler Supply	174
Project 10: Full-Wave Voltage Tripler Supply	180
Project 11: Switchable Full/Half-Voltage Power Supply	183
Project 12: Multi-Output, Add-on Regulator	187
Project 13: Transceiver Power Supply	190
Project 14: Dry Cell Replacement	196
Project 15: 5-Vdc 1-Ampere IC Supply	199
Project 16: Series-Regulated Dual-Polarity Supply	202
Project 17: 300 Vdc Without a Transformer	206
Project 18: 600 Vdc Without a Transformer	209
Project 19: IC-Controlled Variable Voltage Supply	211
Project 20: Solar Power Supply	216
Project 21: Surge Protection for Medium to High Voltage Power Supplies	228
Project 22: High Voltage Dc Power Supply	232
Project 23: Another High Voltage Power Supply	244
Project 24: Dc To Dc Power Supply	252
Project 25: 12-Volt Inverter Circuit	257
Project 26: Free Electricity Power Supply	264
Project 27: 28-Volt Power Supply Using Three Zener Diodes	267
Project 28: Using a Current Meter to Measure Voltage	270
Project 29: A Different Type Of Variable Supply	275
Project 30: Modifying a Low Value Current Meter to Read High Values	277
Project 31: Voltage and Current Readings with a Single Meter	279
Project 32: Altering Secondary Voltage	281

To Ernst P. Schlogvessil, whose patient guidance with this young student provided the future values and drive to excel in realms which far exceeded the immediate subject area.

Other TAB books by the author:

No. 909	*How To Build Metal/Treasure Locators*
No. 996	*Treasure Hunter's Handbook*
No. 1198	*All About AIRGUNS*
No. 1254	*How to Build Hidden, Limited-Space Antennas That Work*
No. 1316	*The Master Handbook of Telephones*
No. 1409	*Build a Personal Earth Station for Worldwide Satellite TV Reception*
No. 1433	*Make Your Own Professional Home Video Recordings*
No. 1487	*The Shortwave Listener's Antenna Handbook*
No. 1496	*The IBM Personal Computer*
No. 1506	*Making Money with Your Microcomputer*
No. 2097	*All About Electric & Hybrid Cars*
No. 2321	*The Joy of Flying*

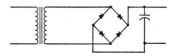

Introduction

A dc power supply is often the first electronic circuit you ever assemble. This small endeavor has led many individuals to years of home brewing electronic devices and even to pursuing an electronics career. Unfortunately, some very fine technicians often consider the workings of a dc power supply too simple and completely beneath them. For this reason, a lot of otherwise knowledgeable individuals know little more than the basics of dc power supplies and their components.

This book is designed to take you from the basics of current flow, electronic components, and circuit assembly through the completion of some fairly complex and certainly exotic power supplies. In the following pages you will find a pleasing blend of theory and practical application. It is not enough just to know how a dc power supply operates. To gain a full understanding of this subject, which can provide a very sound basis for all other electronics pursuits, you must build, test, and operate circuits.

Some of the circuits and projects presented may be familiar, others, although simple, are founded upon practical applications in vogue four decades ago and have all but been forgotten today.

This book not only explains dc power supplies in terminology which can be understood by the neophyte and appreciated by the experienced technician, but also brings you to a point where you will feel familiar and comfortable with the subject matter. The projects are grouped by category; within each category, they are presented in a logical order, starting with the simplest of circuits and progressing to more complex versions.

Start at the beginning of Chapter 1 and absorb and become familiar with theory before you begin building the projects.

This book will educate you in power supply theory and provide you with projects that will demonstrate this theory and also prove practical for many different applications. After you have learned the theory and gained practical building experience, you will be capable of designing and building your own dc power supply projects.

Chapter 1

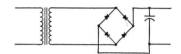

Dc Power Supplies

A dc power supply is any component, device, or circuit that is designed to deliver an output of direct current (dc). It can be a battery, generator, or electronic circuit driven from an alternating current source.

Both alternating current (ac) and direct current (dc) power supplies are instrumental to the working of many different types of electrical and electronic devices. Most electronic devices today require power from a direct current source. Transistors, integrated circuits, and many other components are thought of as direct-current devices. Some electric motors require a direct-current power source, while others are designed to operate from alternating current. Some devices are not selective and will operate with equal efficiency from either ac or dc. These include electric light bulbs, many types of electric heaters, and even the filaments of electron tubes.

Your stereo system, radio, and tape recorder may seem to be powered from alternating current because most are designed to be plugged into a wall outlet that delivers alternating current at a value of approximately 115 volts ac. These devices contain many electronic circuits, most of which are composed of dc devices. Most circuits in these systems will not operate directly from alternating current and are, in fact, driven by direct current. Although ac forms the basic power source, it it converted to direct current. Fortunately, it is far easier to convert alternating current into direct current than the other way around.

VOLTAGE

The term *voltage* describes potential difference. Some people incorrectly refer to voltage as something that flows in a circuit. Voltage does not flow; it is a force which activates a flow of current in a circuit. Voltage is a force and can be likened to a baseball bat that strikes a ball. The movement of the ball can be compared with the flow of current, but the bat is the potential force which causes the movement of the ball. Current and voltage are closely tied to one another, so it is difficult to describe one value without including the other.

A volt is the basic unit for measuring potential difference. There are two basic types of volts, ac and dc. Dc voltage is often supplied by batteries. Ac voltage is obtained from the outlets in a home.

While the volt is a basic unit of potential differences, there are several derivatives which are used to keep numeration to practical levels. The term *millivolt* (mV) is used to describe potential differences that are equivalent to 1/1000 of a volt. An even smaller quantity is the *microvolt*, which is 1/1000 of a millivolt, or 1/1,000,000 of a volt. One *kilovolt* is the equivalent of 1,000 volts. One megavolt is equal to 1,000,000 volts.

When speaking of voltage, it is necessary to know the type of voltage under discussion. In basic electrical circuits, voltage is measured between two points, the positive pole and the negative pole. Contacts from both of these poles must be applied to the circuit to be powered or to the measuring device in order to establish a flow of current. When properly connected to an electronic circuit, or load, a power source will cause current to flow from the negative pole to the positive one.

Dc voltage circuits have fixed poles. The polarity remains constant, in that one contact will always be positive while the other will be negative. Ac power supplies maintain a continual reversal of polarity. During one-half of the ac cycle, one pole will be positive while the other is negative. During the next half of the cycle, the pole that was formerly positive switches to negative while the former negative pole becomes positive. The polarity of the supply undergoes a complete reversal. The rate of polarity change, or alternation, in ac circuits is measured in *hertz* (Hz).

All of us are familiar with alternating current. Our household power is of this type and reverses at a rate of 60 Hz, which means that there are 60 complete cycles in one second. This rate of change is called the ac frequency.

CURRENT

The ampere is the basic unit of current flow. The term flow is quite accurate, as electrons are caused to move through a circuit when a voltage source is connected. Another term which is quite accurate is circuit, because to establish a flow of current, it is necessary to have an electrical path between the positive and negative poles of the voltage source. When this path of flow of circuit is interrupted, current flow ceases.

An ampere is actually the counting of the number of electrons which move past a specific point in a circuit. It is equivalent to approximately six trillion electrons. In practical measurements, the number of electrons is not stipulated by count but is expressed in amperes or in a derivative of this term.

The flow of electrons is always from the negative pole to the positive pole. In dc circuits, the direction of flow is fixed. However, in ac circuits, the current flow continually reverses itself at a rate determined by the frequency of the ac voltage source.

Just as there were many derivatives of the volt, the same applies to the ampere. The milliampere is equal to 1/1000 of an ampere and is a term which is often used to describe the current values found in solid-state electronic circuits. The microampere is equal to 1/1,000,000 of an ampere, or 1/1000 of a milliampere, and is used to describe the current values of extremely low powered electronic circuits.

Whereas the measurement of voltage is accomplished by placing a multimeter probe at the positive and negative poles of the voltage source, current is measured by placing a ammeter or milliammeter directly within the current path.

To better understand the differences between voltage and current, think of current as matter and voltage as force. If you throw a rock at a target, the force with which this object is tossed can be compared to voltage. The rock itself is the current. Obviously, the final impact at the target will be dependent upon the size of the rock as well as the force with which it is thrown. Likening this to an electronic circuit, the effect current flow has will depend upon the amount of current and the voltage behind it.

The force or voltage behind alternating current varies periodically. Figure 1-1 shows a plot of a standard ac waveform which might be seen on an oscilloscope attached to your household line. The starting voltage point is zero, but this value immediately begins to rise to a peak value. Past the peak, the voltage value begins to decay again and returns to zero. This completes the first half of the ac

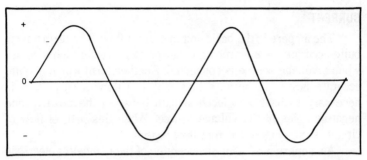

Fig. 1-1. Graph of an ac sine wave.

cycle. Notice that the voltage rise and fall in this portion is all of a positive polarity. After the first half of the ac wave has been completed, the curve continues on through zero volts and completes a mirror image of the previous waveform, but this time at a negative polarity. After the bottom of the negative dip is passed, the voltage again returns to zero. At this point, one full alternating current wave has been completed. With household current, one waveform or cycle occurs every 1/60th of a second. This means that 60 complete cycles occur in every one-second interval. Thus, the designation of household current in the United States as 60 cycles per second (cps) or 60 Hz is given.

Ac lines in most homes are rated at 115 volts. But notice that the voltage value starts at zero, climbs to a peak value of about 150 positive volts, returns to zero again, and dips to a peak negative value of 150 volts before completing the cycle. The 115-volt value is actually an effective value. It has the same heating effect as 115 dc volts. The 115-volt value is often called the *root-mean-square* value. It is approximately equal to 70 percent of the total waveform.

The ac waveform produced by your local electric company is said to be *sinusoidal*. This means that the waveform produced by one half cycle is a mirror image of that produced in the other. Many other types of alternating current waveforms are produced in different kinds of electronic circuits, but these do not apply to this discussion.

The purpose of a dc power supply is to change alternating current into direct current. The latter is defined as a current flow of one polarity only. Alternating current changes polarity at regular intervals.

While the conversion from ac to dc is extremely simple from a builder's standpoint, the actual process is a little more complicated. Rectification is a term used to describe the conversion of alternating current to direct current. Electronic devices which perform this function are known as rectifiers.

Figure 1-2 shows the previous ac waveform after rectification has taken place. The positive half cycle is allowed to pass, but the negative portion is blocked. The blocked portion is indicated by dotted lines. Later, we will go into more depth on the components involved in the rectification process. It is enough now to know that rectification involves the passing of current of a single polarity.

The rectified waveform shown in Fig. 1-2 matches the positive portion of the ac waveform in every detail. The output is now true dc or direct current, in that the definition of direct current is met, since we are now dealing with a current that is positive in respect to ground or zero. If we had used the bottom or negative half of the ac cycle, this would still be direct current with the output being negative in respect to ground.

While this is true direct current, it is not what is known as pure dc. Dc power sources for most electronic circuits must usually be of a single amplitude or voltage value. The waveform in Fig. 1-2 represents *pulsating* dc. This means that while the polarity does not change, the voltage value does. The direct current shown varies from zero to about 150 volts.

To arrive at pure dc, a filter circuit is required which serves to smooth out the ac ripple component in the dc output. Once the current has passed through the filter circuit, it will be of one amplitude which will be roughly equivalent to the ac peak value or 150 volts. The actual value will depend upon the type of filter circuit used and the overall current demand. As discussed here, an ac

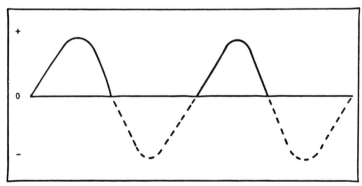

Fig. 1-2. During rectification, one-half of the sine wave is conducted, while the other half is suppressed. The dotted lines indicate the suppressed portion of the wave. Since only half of the sine wave is conducted, this is called half-wave rectification and its pulsating dc output has a frequency of 60 hertz.

voltage with an RMS value of 115 volts will result in a dc output of about 150 volts after rectification and filtering have taken place. Figure 1-3 shows a graphic representation of a pure dc waveform.

The rectification process just discussed acts upon only one half of the ac waveform. The remaining portion is blocked from passing through the circuit. This process is called half-wave rectification. This is not desirable in many applications due to the more stringent filtering requirements needed to arrive at a pure dc output. Full-wave rectifier circuits are easily constructed which act upon both the negative and positive portions of the ac sine wave. Here, the negative portion of the cycle is also conducted but is electronically switched into the positive polarity area. The output from a full-wave rectifier is shown graphically in Fig. 1-4. Notice that there are twice as many current pulses as were obtained with the half-wave circuit. This graph represents the same time interval as before, or 1/60th of a second. We are still dealing with pulsating dc, but notice that there are twice as many voltage pulses as before. Therefore, the pulse frequency of a full-wave rectifier circuit is twice that of a half-wave design. Higher pulse frequencies are more easily filtered into a pure dc output than are those of a lower frequency rate.

This has been a very basic discussion of the rectification process and some of the more technical aspects have been simplified for a better understanding. As you continue on in this text, the operation of the electronic circuits used to accomplish this and other processes will become even more clear.

Dc power supplies are available in many different types. Some are designed to deliver a small amount of current at low voltages, while others may deliver massive amounts of current at the same

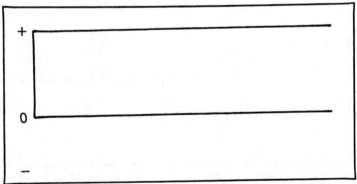

Fig. 1-3. Representation of pure dc. Note that the voltage quickly swings from 0 to a peak positive value and then holds this value for the entire duration of operation.

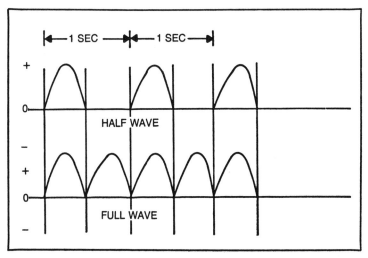

Fig. 1-4. Graphic representations of pulsating dc waveforms from half-wave and full-wave rectifier circuits.

potential values. Still others may supply high voltage potentials at moderate to low current levels. Some may be very complex in nature, while others will be extremely simple. The demands placed upon the power supply will dictate the degree of complexity as well as the component ratings of each design.

SUMMARY

Dc power supplies are used in almost every phase of electronics. Usually, the dc will be derived from rectified alternating current which is commonly available as a major source of electrical power throughout the United States. Other dc supplies will consist of storage batteries, solar cells, and mechanical generators.

When converting from alternating current to direct current, it is necessary to arrive at an output which is of a single polarity with respect to ground and also of a stable single voltage value. The rectification process is only one of many steps which are accomplished by electronic components and the circuits they form. But this is the first step of the conversion process which will end in a complex circuit which delivers the voltage and current you desire.

Chapter 2

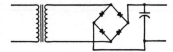

The Basic Dc Power Supply

No matter how complicated a dc power supply circuit may be, it can probably be broken down into several basic sections common to all power supplies. For most practical applications, dc power supplies will be driven from the ac supply in your home or shop. While voltage levels will vary from area to area, household current is usually rated at a value of 120 and 240 volts. Other values may be given of 115, 117, or 120 volts, but all of these refer to the standard household mains which usually fall between 110 and 120 volts ac in most areas of the United States.

Basically, the ac supply is connected at the primary input of the power transformer. The transformer output can range anywhere from about 2 volts ac to many thousands of volts, depending upon the secondary winding. The output voltage from the transformer secondary will determine the final dc output voltage from the power supply, but only after several other electronic circuits are added.

RECTIFICATION

Rectification is described as the changing of an alternating current (ac) to a unidirectional or direct current (dc). The normal PN junction diode is well suited for this purpose, as it conducts very heavily when forward biased (low resistance direction) and only slightly when reverse biased (high resistance direction).

The use of PN junction rectifiers in the design of power supplies for electronic equipment is increasing. Reasons for this include these characteristics: no requirement for filament power, immediate operation without need for warm-up, low internal voltage drop substantially independent of load current, low operating temperature, and generally small physical size. PN junction rectifiers are particularly well-adapted for use in the power supplies of portable and small electronic equipment where weight and space are important considerations.

Semiconductor materials treated to form PN junctions are used extensively in electronic circuitry. Variations in doping agent concentrations and physical size of the substrate produce diodes which are suited for different applications. There are signal diodes, rectifiers, zener diodes, reference diodes, varactors, and others.

DIODES

Pictorial representations of various diodes are shown in Fig. 2-1. This is but a very limited representation of the wide assortment in case design. However, the shape of characteristic curves of these diodes is very similar. Primarily, current and voltage limits and relationships are different. Figure 2-2 shows a typical curve of a junction diode. The graph shows two different kinds of bias. Bias in the PN junction is the difference in potential between the anode (P material) and the cathode (N material). Forward bias is the application of a voltage between N and P material, where the P material is positive with respect to the N material. When the P material becomes negative with respect to the N material, the junction is reverse biased. Application of greater and greater amounts of forward bias causes more and more forward current until the power handling capability of the diode is exceeded, unless limited by external circuitry. Small amounts of forward bias cause very little current flow until the internal barrier potential is overcome. The potential difference varies from diode to diode but is usually no more than a few tenths of a volt. Reverse bias produces a very small

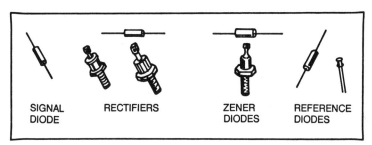

Fig. 2-1. Pictorial representations of various junction diodes.

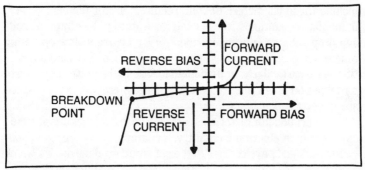

Fig. 2-2. Junction diode characteristic curve.

amount of reverse current until the breakdown point is reached. Then, an increase in reverse bias will cause a large increase in reverse current. Therefore, if breakdown is not exceeded, the ratio of forward current to reverse current is large; for example, milliamperes to microamperes, or amperes to milliamperes. Changes in temperature may cause alterations in the characteristic curve, such as: slope of curve at any point, breakdown point, amount of reverse current, etc.

Diode Specifications

There are many specifications listed in various manufacturers' specification sheets and in semiconductor data manuals. Descriptions of various diode types and their more important electrical characteristics are necessary for troubleshooting and design.

Rectifier Diodes

Rectifier diodes are used primarily in power supplies. These diodes are usually of the silicon type because of this material's inherent reliability and higher overall performance compared to other materials. Silicon allows higher forward conductance, lower reverse leakage current, and operation at higher temperatures compared to other materials.

The major electrical characteristics of rectifier diodes are listed below:

Dc Blocking Voltage (V_R). Maximum reverse dc voltage which will not cause breakdown.

Average Forward Voltage Drop (V_F). Average forward voltage drop across the rectifier given at a specified forward current and temperature, usually specified for rectified forward current at 60 Hz.

Average Rectifier Forward Current (I_F). Average rectified forward current at a specified temperature, usually at 60 Hz with a resistive load. The temperature is normally specified for a range, typically −65 to +175 degrees centigrade.

Average Reverse Current (I_R). Average reverse current at a specified temperature, usually at 60 Hz.

Peak Surge Current (I_{SURGE}). Peak current specified for a given number of cycles or portion of a cycle. For example, ½ cycle at 60 Hz.

Zener Diodes

The zener diode is unique compared to other diodes, in that it is designed to operate reverse biased in the avalanche or breakdown region. The device is used as a regulator, clipper, coupling device, and in other functions in computer systems.

The major electrical characteristics of zener diodes are:

Nominal Zener Breakdown ($V_{Z(NOM)}$). Sometimes a $V_{Z(MAX)}$ and $V_{Z(MIN)}$ are used to set absolute limits between which breakdown will occur.

Maximum Power Dissipation (P_D). Maximum power the device is capable of handling. Since voltage is a constant, there is a corresponding current maximum (I_{ZM}).

Schematic diagrams of the zener are shown in Fig. 2-3. Zener current flows in the direction of the arrow. In many schematics, a distinction is not made for this diode, and a signal diode symbol is used.

PN JUNCTION DIODE RECTIFIER CIRCUITS

Figure 2-4 is a block diagram of a power supply showing an ac input to and a dc output from a block labeled positive power supply and filter network. Although this figure shows a power supply that provides a unidirectional current which causes a positive voltage

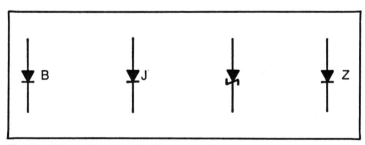

Fig. 2-3. Schematic symbols for the zener diode.

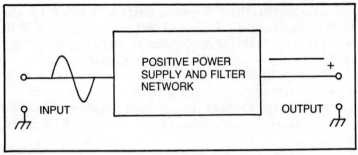

Fig. 2-4. Positive voltage output from an ac input.

output, it might well be designed to furnish a negative voltage output.

The question of why a change of ac to dc is necessary probably arises. The answer, simply, is that for proper operation, many electronic circuits depend on direct current. The PN junction diode conducts more easily in one direction than in the other. Transistors and electron tubes are also unidirectional and a constantly alternating source voltage would be undesirable.

Before describing how an ac input is converted into a dc output, the term load as it applies to power supplies must be understood. Load is the current supplied to the power-consuming device or devices connected to the power supply. The power-consuming device needs voltage and current as supplied by the power supply. The power-consuming device may be a simple resistor or one or more electronic circuits using resistors, capacitors, coils, and active devices.

HALF-WAVE RECTIFIER

Figure 2-5 shows the PN junction diode functioning as a half-wave rectifier. A half-wave rectifier is one that uses only half of the input cycle to produce an output.

The induced voltage across L2 (the transformer secondary) will be as shown in this figure. The dots on the transformer indicate points of the same polarity. During that portion of the input cycle which is going positive (solid line), CR1, the PN junction diode, will be forward biased and current will flow through the circuit. L2, acting as the source voltage, will have current flowing from the top to the bottom. This current then flows up through R_L, causing a voltage drop across R_L equal to the value of current flowing times the value of R_L. This voltage drop will be positive at the top of R_L with respect to its other side, and the output will therefore be a positive voltage with respect to ground. It is common practice for the end of a resistor receiving current to be given a sign representing a negative polarity of voltage, and the end of the resistor through which current leaves is assigned a positive polarity of voltage. The voltage drop across R_L, plus the voltage drops across the conducting diode and L2, will equal the applied voltage. Although the output voltage will nearly equal the peak input voltage, it cannot reach this value due to the voltage drops, no matter how small, across CR1 and L2.

The negative half cycle of the input is illustrated by the broken line. When the negative half cycle is felt on CR1, the PN junction diode is reverse biased. The reverse current will be very small, but it will exist. The voltage resulting from the reverse current, as shown below the line in the output is exaggerated in this figure to bring out the point of its existence. Although only one cycle of input is shown here, it should be realized that the action described continually repeats itself as long as there is an output.

By reversing the diode connection in this figure (having the anode on the right instead of the left), the output would now become a negative voltage. The current would be going from the top of R_L toward the bottom, making the output at the top of R_L negative in respect to the bottom or ground side.

The same negative output can be obtained from this figure if the reference point (ground) is changed from the bottom (where it is shown) to the top or cathode-connected end of the resistor. The bottom of R_L is shown as being negative in respect to the top, and reading the output voltage from the hot side of the resistor to ground would result in a negative voltage output.

The half-wave rectifier will normally indicate improper functioning in one of two manners: there is no output or the output is low. The no output condition can be caused by no input; i.e., the fuse has blown, the transformer primary or secondary winding is open, the PN junction diode is open, or the load is open.

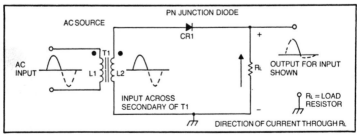

Fig. 2-5. Positive voltage output half-wave PN junction diode rectifier.

13

The low output condition might be caused by an aged diode. A check of both forward and reverse resistance of the diode may reveal the condition of the diode. Low output can be the result of an increased forward resistance or a decreased reverse resistance of the diode.

It is necessary to check the ac input voltage to see if it is of the correct value. A low input voltage will result in a low output voltage. A check of the transformer secondary voltage should also be made to see if it is of the correct value also, as a low secondary voltage will also result in a low output voltage.

By removing the input voltage, you can make resistance checks of the components. Is the primary open? Is the secondary open? Has the diode increased in forward resistance value or decreased in reverse resistance value? Is the diode open? Has the load resistance become shorted? Do the components show signs of excessive heat dissipation? Have they become discolored? Does energizing the circuit and putting an ammeter in series with the load make the load current excessive?

You can answer all these questions when troubleshooting the half-wave rectifier. If you discover a problem in the rectifier, then determine if the cause is a local one (in the rectifier itself). Or, is it due to some changes in the following circuitry, such as the power supply filter components or changing the load impedance? While it is important that the problem be repaired, elimination of the cause is of even greater importance.

FULL-WAVE RECTIFIER

The PN junction diode works just as well in a full-wave rectifier circuit, as shown in Fig. 2-6. The circuit shown has a negative voltage output. However, it might just as well have a positive voltage output. This can be accomplished by either changing the reference point (ground side of R_L) or by reversing the diodes in the circuit.

The ac input is felt across the secondary winding of T1. This winding is center tapped as shown (the center of the secondary is at ground potential). When the polarity is such that the top of T1 secondary is negative, the bottom is positive. At this time, the center tap, as shown, has two polarities, positive with respect to the top half of the winding, and negative with respect to the bottom half of the winding. When the secondary winding is positive at the top, the bottom is negative and the center tap is negative with respect to the top and positive with respect to the bottom. What is the polarity

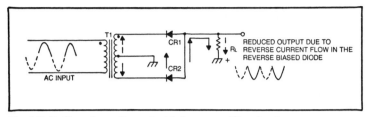

Fig. 2-6. Unfiltered negative output full-wave rectifier circuit.

of the reference point (ground)? The answer must be in terms of the "with respect to" portion of the statement. For each alternation of the input, one of the diodes will be forward biased and the other one reverse biased.

For ease of explanation, the negative alternation will be considered when the rectifier current is initially energized by the ac source. CR1 will be forward biased (negative voltage felt on its cathode) and CR2 will be reverse biased (a positive voltage felt on its cathode). Therefore, the top of T1 secondary must be negative with respect to the bottom. When forward bias is applied to CR1, it conducts heavily from cathode to anode (dashed arrow), down through R1 (this current flow creates a voltage drop across R_L), and negative at the top with respect to the bottom or ground side of R_L. The current passing through R_L is returned to CR1 by going through the grounded center tap and up the upper section of the center tapped secondary winding of T1. This completes the first alternation of the input cycle. The second alternation of the input now is of such polarity as to forward bias CR2 (a negative voltage at the bottom of T1 secondary winding with respect to ground). CR1 is now reverse biased. CR2 conducts, current moves in the same direction through R_L (solid arrow) top to bottom, and back through the lower half of the center-tapped secondary to CR2. You may wonder why current does not flow from the anode of one of the diodes through the anode to cathode of the other diode. The answer is simple. It does. However, current flow through a reverse biased diode is very slight due to the high resistance of the diode when in this condition. This rectifier has a slightly reduced output, as shown in Fig. 2-6, because of the reverse current flow.

As you can see in the output waveform of this figure, there are two pulses of dc out for every cycle of ac in. This is full-wave rectification. Current flow through R_L is in the same direction, no matter which diode is conducting. The positive going alternation of the input allows one diode to be forward biased and the negative going alternation of the input allows the other diode to be forward biased. The output for the full-wave rectifier shown is a negative voltage measured from the top of R_L to ground.

As in the half-wave rectifier, there can be two indications of trouble: no output or low output. No output conditions are indications of no input, shorted load circuits, open primary winding, open or shorted secondary winding, or defective diodes. Low output conditions are possible indications of aging diodes, open diodes, or opens in either half of the secondary winding (allowing the circuit to act as a half-wave rectifier).

The method for troubleshooting the full-wave rectifier is the same as that used for the half-wave rectifier. Check voltages of both primary and secondary windings, check current flow, and when the circuit is deenergized, take resistance measurements. Shorted turns in the secondary windings give a lower voltage output and possibly shorted turns in the primary winding will produce a lower voltage input. (Shorted turns are hard to detect with an ohmmeter. They are more easily detected by taking a voltage reading across various terminals of the energized transformer.)

BRIDGE RECTIFIER

Figure 2-7 shows a PN junction used in a bridge rectifier circuit. This circuit is capable of producing a positive output voltage. When the ac input is applied across the secondary winding of T1, it will forward bias diodes CR1 and CR3, or CR2 and CR4. When the top of the transformer is positive with respect to the bottom, as illustrated in this figure by the designation number 1, both CR1 and CR2 will feel this positive voltage. CR1 will have a positve voltage on its cathode, a reverse bias condition, and CR2 will have a positive voltage on its anode, a forward bias condition. At this same time, the bottom of the secondary winding will be negative with respect to the top, placing a negative voltage on the anode of CR2 (a reverse bias condition) and on the cathode of CR4 (a forward bias condition).

During the half cycle of the input designated by the number 1 in this figure, CR2 and CR4 are forward biased and will therefore conduct heavily. The conducting path is shown by the solid arrows, from the source (the secondary winding of T1) through CR4 to ground, up through R_L, making the top of R_L positive with respect to the ground end, to the junction of CR2 and CR3. CR2, being forward biased, offers the path of least resistance to current flow and this is the path current will take to get back to the source.

During the alternation designated by the number 2 in this figure, as indicated by the dashed arrows, the top of the secondary

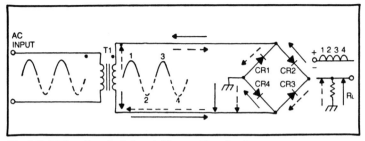

Fig. 2-7. PN junction diode in a bridge rectifier circuit.

winding is going negative while the bottom is going positive. The negative voltage at the top is felt by both CR1 and CR2, forward biasing CR1 and reverse biasing CR2. The positive voltage on the bottom of T1 secondary is felt by CR3 and CR4, forward biasing CR3 and reverse biasing CR4. Current flow starting at the source (T1 secondary winding) is through CR1 to ground, up through R_L (this is the same direction as it was when CR2 and CR4 were conducting, making the top of R_L positive with respect to its ground end), to the junction of CR2 and CR3. This time, CR3 is forward biased and offers the least opposition to current flow, and current takes this path to return to its source.

Notice that the diodes in the bridge circuit operate in pairs. First one pair, CR1 and CR3, operate heavily; then the other pair, CR2 and CR4, conduct heavily. The output waveform shows one output pulse for every half cycle of the input, or two pulses out for every cycle in. This is the same as for the full-wave rectifier circuit explained earlier.

The bridge circuit will also indicate a malfunction in one of two manners: it has no output or a low output. The causes for both conditions are the same as they were for the half- or full-wave rectifier. If any one of the diodes opens, the circuit will act as a half-wave rectifier with a resultant lower output voltage.

JUNCTION DIODE CONSIDERATIONS

The junction diode has four important ratings that must be taken into consideration when designing a power supply. They are the maximum:

- Average forward current.
- Repetitive reverse voltage.
- Surge current.
- Repetitive forward current.

These ratings are important when it becomes necessary to troubleshoot a power supply or when selecting junction diodes for replacement when the desired one is not readily available.

The maximum average forward current is the maximum amount of average current that can be permitted to flow in the forward direction. This rating is usually given for a specified ambient temperature and should not be exceeded for any length of time, as damage to the diode will occur. The maximum repetitive reverse voltage is that value of reverse bias voltage that can be applied to the diode without causing it to break down.

The maximum surge current is that amount of current allowed to flow in the forward direction in non-repetitive pulses. Current should not be allowed to exceed this value at any time and should only equal this value for a period not to exceed one cycle of the input. The maximum repetitive forward current is the maximum value of current that may flow in the forward direction in repetitive pulses.

All of the ratings mentioned above are subject to change with temperature variations. If the temperature increases, the ratings given on the specification sheet should all be lowered or damage to the diode will result.

POWER TRANSFORMERS

Power transformers are used in power supply circuits because of the efficiency and ease with which they transfer energy. The power transformer is capable of receiving a voltage at one level and delivering it at the same level, some higher level, or some lower level. Transformers that convert voltage to a high level are called step-up transformers. Those that convert voltage to a lower level are called step-down transformers, and those that provide the same output voltage level as the applied input level are one-to-one transformers. The references in all cases are to voltage levels.

Even though the coils in power transformers are wound on the same material, the separate windings are each insulated and are therefore electrically isolated from one another. The source of energy for the primary is thus isolated from the secondary and any associated circuitry. The efficiency of power transformers is very high. Approximately 90 percent of the input power is usable in the output.

A quick review of voltage, current, and power relationships is in order at this point. The voltage induced into the secondary winding of any given transformer is determined by the ratio of the number of turns in the primary winding to the number of turns in the secondary winding and the amount of voltage applied to the primary. For every volt per turn of the primary winding, there will be a volt per turn in the secondary winding. For example, if the primary has ten windings (turns) and each winding has ten volts, the entire primary has 100 volts applied.

The secondary winding in this example has only five windings. Since the number of volts per turn must be the same for primary and secondary windings, we find that the secondary winding has only 50 volts (5 windings and 10 volts per winding = 50 volts). This is an example of a step-down transformer. If, instead of 5 windings in the secondary. The voltage induced in the secondary is determined by would be 150 volts (15 windings at 10 volts per winding) and the transformer would be a step-up transformer.

From this information, a simple relationship can be derived. The voltage in the primary (E_p) is equal to the number of volts per turn times the number of turns (N_p) in the primary; and the voltage in the secondary (E_s) is equal to the number of turns (N_s) in the secondary. The voltare induced in the secondary is determined by the ratio $E_p/E_s = N_p/N_s$. It is now a simple matter to calculate any missing value when the other three are given.

Primary power ($E_p I_p$) would equal secondary power ($E_s I_s$) in an ideal transformer (the power taken from the source would be equal to the power delivered to the load). The power transformer is an efficient element, but it is not 100 percent efficient. The reasons for the losses will not be discussed in detail here, but the types of losses encountered in power transformers are: (1) Copper or I^2R Losses, (2) Eddy Current Losses, and (3) Hysteresis Loss. Due to these losses, the efficiency of the power transformer encountered in electronic equipments is about 90 percent.

The efficiency of the power transformer is found by the formula % of Efficiency = $P_{out}/P_{in} \times 100$. The current ratio is the reverse of the voltage or turns ratio ($I_s I_p$). This means that in a step-up transformer, the current flow in the secondary is less than that in the primary. In a step-down transformer, the current flow in the secondary is greater than that in the primary. When the voltage in the primary remains the same (no changes) and the voltage in the secondary increases, the current in the secondary decreases (power out = power in). This is true because the flux lines developed in a transformer core are proportional to the ampere turns of the associated windings (since the flux is the same for both windings, the ampere turns must be the same for both) an $N_p I_p$ equals $N_s I_s$. Since

$$r = \frac{E_{RMS}}{E_{dc}}$$

Fig. 2-8. Formula used to determine ripple factor.

N_p/N_s equals E_p/E_s, it follows that $E_p I_p$ equals $E_s I_s$. It has been noted that secondary voltage may be higher or lower than the primary voltage. This being true, then $E_p/E_s\ I_s I_p$ or the primary, and secondary currents are inverse to the voltage.

POWER SUPPLY FILTERS

The operation of most electronic circuits is dependent upon a direct current source. It has been illustrated how alternating current can be changed into a pulsating current; that is, a current that is always positive or negative with respect to ground although it is not of a steady value. This means it has ripple.

Ripple can be defined as the departure of the waveform of a rectifier from pure dc. It is the amplitude excursions, positive and negative, of a waveform from the pure dc value (the alternating component of the rectifier voltage). Ripple contains two factors which must be considered: frequency and amplitude. Ripple frequency in the rectifiers that have been presented is either the same as line frequency (for the half-wave rectifier) or twice the line frequency (for the full-wave rectifier).

In the half-wave rectifier, one pulse of dc output was generated for one cycle of ac input. The ripple frequency is the same as the input frequency. In the full-wave rectifiers (center tapped and bridge), two pulses of dc output were produced for each cycle of ac input. The ripple frequency is twice that of the line frequency. With a 60 hertz input frequency, there will be a 60 hertz ripple frequency in the output of a half-wave rectifier and a 120 hertz ripple frequency in the output of the full-wave rectifier.

The amplitude of the ripples in the output of a rectifier circuit will provide a measure of the effectiveness of the filter being used, or the ripple factor. The ripple factor is defined as the ratio of the RMS value of the ac component to the average dc value. This is shown in the formula in Fig. 2-8. The lower the ripple factor, the more effective the filter. The term percent of ripple may be used. This is different from the ripple factor only because the figure arrived at in the ripple factor formula is multiplied by 100 to give a percent figure. This is shown in Fig. 2-9. In both formulas, E_{RMS} is the RMS value of ripple voltage and E_{dc} is the dc value (average value) of the output voltage.

$$\% \text{ Ripple} = \frac{E_{RMS}}{E_{dc}} \times 100$$

Fig. 2-9. In this formula, the factor arrived at in the previous formula is merely multiplied by 100 to arrive at a percentage.

Filter circuits used in power supplies are usually low pass filters. (A low pass filter is a network which passes all frequencies below a specified frequency with little or no loss but is highly discriminate against all higher frequencies.) The filtering is done through the use of resistors or inductors and capacitors. The purpose of power supply filters is to smooth out the ripple contained in the pulses of dc obtained from the rectifier circuit while increasing the average output voltage or current.

Filter circuits used in power supplies are of two general types, capacitor input and choke input. There are several combinations that may be used, although they are referred to by different names (Pi, RC, L section, etc.). The closest element electrically to the rectifier determines the basic type of filter being used.

Figure 2-10 depicts the basic types of filters. In A, a capacitor shunts the load resistor, therein bypassing the majority of ripple

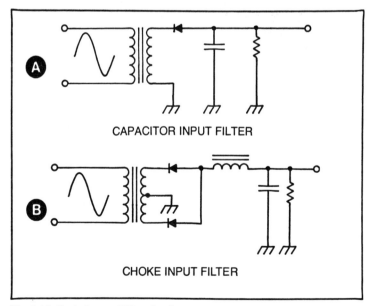

Fig. 2-10. Basic types of filter circuits.

21

current which passes through the series elements. In B, an inductor (choke) in series with the load resistor opposes any change in current in the circuit. The capacitor input filter will keep the output voltage at a higher level compared to a choke input. The choke input will provide a steadier current under changing load conditions. Therefore, you would use a capacitor input filter where voltage is a prime factor, and you would use the choke input filter where a steady flow of current is required.

Capacitor Input Filter

First, an analysis will be made of the simple capacitor input filter depicted in Fig. 2-11A. The output of the rectifier without filtering is shown in Fig. 2-11B and the output after filtering is illustrated in Fig. 2-11C. Without the capacitor, the output across R_L will be pulses. The average value of these pulses would be the E_{dc} output of the rectifier.

With the addition of the capacitor, the majority of the pulses changes are bypassed through the capacitor and around R_L. As the first pulses appears across the capacitor, changing it from negative to positive, bottom to top, the peak voltage is developed across the capacitor. When the first half cycle has reached its peak and starts its negative going excursion, the capacitor will start to discharge through R_1, maintaining the current through R_L in its original direction and thereby holding the voltage across R_L at a higher value than its unfiltered load. Before the capacitor can fully discharge, the positive excursion of the next half cycle is nearing its peak, recharging the capacitor. As the pulse again starts to go negative, the capacitor starts to discharge once again. The positive going excursion of the next half cycle comes in and recharges the capacitor. This action continues as long as the circuit is in operation.

The charge path for the capacitor is through the transformer secondary and the conducting diodes, and the discharge path is through the load resistor. The reactance of the capacitor at the line

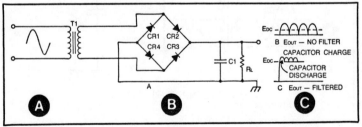

Fig. 2-11. Low voltage power supply with simple capacitor filter.

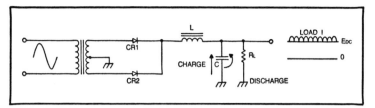

Fig. 2-12. L-section (choke input) filter showing current waveforms.

frequency is small compared to R_L, which allows the changes to bypass R_L and effectively, only pure dc appears across R_L.

This illustrates the use of an RC time constant. If the value of C1 or R_L were such that the discharge time was the same or less than that of the charge time, we would have no filtering action. The larger the values of C1 and R_L, the longer the time constant and the lower the ripple factor. The charge time of the capacitor is the RC values of the capacitor, the conducting diodes, and the transformer secondary. The impedance offered by these elements is very small when compared to the impedance in the discharge path of the capacitor (the value of R_L). The output voltage is practically the peak value of the input voltage. This circuit provides very good filtering action for low currents, but results in little filtering in higher current power supplies due to the smaller resistance of the load.

Choke Input Filter

The next filter to be discussed is the choke input filter or the L-section filter. Figure 2-12 shows this filter and the resultant output of the rectifier after filtering has taken place. The series inductor, L(choke) will oppose rapid changes in current. The output voltage of this filter is less than that of the capacitor input filter, since the choke is in series with the output impedance. The parallel combination of R_L and C in connection with L smoothes out the peaks of the pulses and results in a steady, although reduced, output.

The inductance chokes off the peaks of the alternating components of the rectifier waveform and the dc voltage is the average or dc value of the rectified wave. The choke input filter allows a continuous flow of current from the rectifier diodes rather than the pulsating current flow as seen in the capacitor input filter. The X_L of the choke reduces the ripple voltage by opposing any change in current during either the rapid increases in current during the positive excursions of the pulses or decreases in current during the negative excursions. This keeps a steady current flowing to the load throughout the entire cycle. The pulsating voltage developed across the capacitor is maintained at a relatively constant value approaching the average value of input voltage because of this steady current flow.

Multiple Section Choke Input Filter

The filtering action provided by the choke input filter can be enhanced using more than one such section. Figure 2-13 shows two sections with representative waveforms approximating the shape of the voltage with respect to ground at different points in the filter networks.

While this figure shows two choke input sections being used as a multiple section filter, more sections may be added as desired. While the multiple section filter does reduce the ripple content (and they are found in applications where only a minimum ripple content can be tolerated in the output voltage), they also result in reduced regulation. The additional sections add more resistance in series with the power supply, which results in increased voltage variations in the output when the load current varies.

Pi Filter

A filter, called the pi filter because of its resemblance to the Greek letter π (prounced pi), is a combination of the simple capacitor input filter and the choke input filter. This filter is shown in Fig. 2-14.

The resistor, R, is known as a bleeder resistor and is found in practically all power supplies. The purpose of this resistor is twofold. When the equipment has been working and is then turned off, it provides a discharge path for the capacitor, preventing a possible shock to maintenance personnel; and it also provides a fixed load, no matter what equipment is connected to the power supply. It is also possible to use this resistor as a voltage dividing network through the use of appropriate taps.

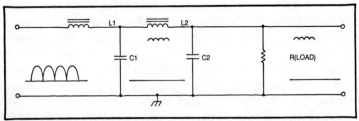

Fig. 2-13. Multiple section choke input filter with representative waveforms.

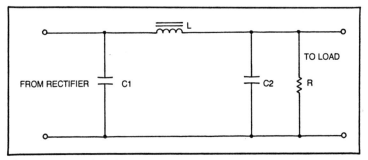

Fig. 2-14. The basic Pi filter.

The pi filter is basically a capacitor input filter with the addition of an L-section filter. The majority of the filtering action takes place across C1, which charges through the conducting diode(s) and discharges through R, L, and C2. As in the simple capacitor input filter, the charge time is very fast compared to the discharge time. The inductor smooths out the peaks of the current pulses felt across C2, thereby providing additional filtering action. The voltage across C2, since C2 is in parallel with the output, is the output voltage of the power supply. Although the voltage output is lower in this filter than it would be if taken across C1 and the load, the amount of ripple is greatly reduced.

Even though C1 will charge to the peak voltage of the input when the diodes are conducting and discharge through R when they are cut off, the inductor is also in the discharge path and opposes any changes in load current. The voltage dividing action of L and C2 is responsible for the lower output voltage in the pi filter when compared to the voltage available across C1.

In the previously referenced Fig. 2-14, the charge path for both C1 and C2 is through the transformer secondary, through the capacitor, and in the case of C2, through L. Both charge paths are through the conducting diode in the rectifier. However, the discharge path for C1 is through R and L, while the discharge path for C2 is through R only. How fast the input capacitor C1 discharges is mainly determined by the ohmic value of R. The discharge time of the capacitors is directly proportional to the value of R. If C1 has very little chance to discharge, the output voltage will be high. For lower values of R1, the discharge rate is faster and the output voltage will decrease. With a lower value of resistance, the current will be greater and the capacitor will discharge further. The E output is the average value of dc; and the faster the discharge time, the lower the average value of dc and the lower the E out.

Rc Capacitor Input Filter

While the pi filter previously discussed had an inductor placed between two capacitors, the inductor can be replaced by a resistor, as shown in Fig. 2-15. The main difference in operation between this pi filter and the one previously discussed is the reaction of an inductor to ac when compared to the resistor. In the former filter, the combination of the reactances of L and C2 to ac was such as to provide better filtering, giving a relatively smooth dc output.

In Fig. 2-15, both the ac and dc components of rectified current pass through R1. The output voltage is reduced due to the voltage drop across R1; and the higher the current, the greater this voltage drop. This filter is effective in high voltage, low current applications. As in choke input filters, the capacitor input filters shown may be multiplied; i.e., identical sections may be added in series. The choice of a filter for a particular use is a design problem, but you should understand the purpose and operation of filters because of their importance to the proper operation of equipment utilizing a power supply.

BLEEDER RESISTOR VOLTAGE DIVIDER

Figure 2-16 shows how the bleeder resistor may function as a voltage divider network. Terminal 3 is grounded. Current is flowing as indicated from the bottom to the top, making terminal 4 negative with respect to ground (terminal 3). Terminals 1 and 2, on the other hand, are positive with respect to ground.

Since collector voltage from NPN and PNP transistors used in amplifier circuits needs positive and negative values, respectively, a voltage dividing network such as that shown in Fig. 2-16 is typical. At point X, 50 mA (milliamperes) of current enters the junction of R3 and terminal 4 (load A). At point X, this current divides, 40 mA

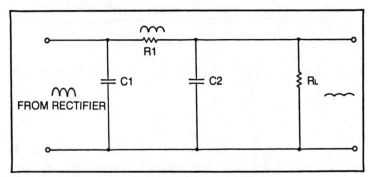

Fig. 2-15. Capacitor input filter and associated waveforms.

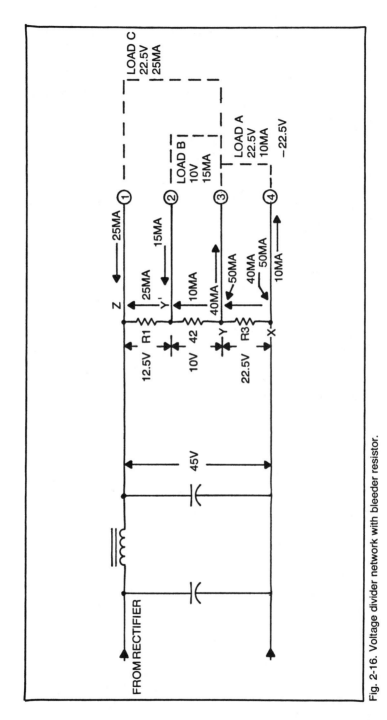

Fig. 2-16. Voltage divider network with bleeder resistor.

flows through R3, and 10 mA through load A. They both have a voltage drop of 22.5 volts across them.

At point Y, the current again divides into the parallel paths of R2, load B, and load C. As indicated, 40 mA flows through loads B, and C, and 19 mA flows up through R2. The current through R2 causes a voltage drop of 10 volts. The top of R2 is positive with 10 volts with respect to the bottom, or ground side, of R2. Load B, being in parallel with R2, is also a positive 10 volts, although there are 15 mA of current flowing through load B. It should be evident that the value of R2 is greater than the impedance of load B by a ratio of 1.5 to 1. If it isn't evident, work it out using the formula in Fig. 2-17, since you have the values of both E and I. At point Y', the currents through R2 and load B rejoin at terminal 2 and flow up through R1. The 25 mA flowing through R1 makes the voltage drop across R1 a positive 12.5 volts with respect to its lower end. The voltage measured from point Z to ground, however, will be 22.5 volts, since the voltage drops across R1 (12.5 volts) and R2 (10 volts) are between point Z and ground. (The voltage across load C is in parallel with the series combination of R1 and R2, with load B in parallel with R2.) At point 2, the 25 mA flowing in load C combines with the 25 mA in R1, which satisfies Kirchhoff's current law.

Figure 2-18 shows this division and joining of current in line form, with the arrowheads indicating the direction of current. While it may seem that current is flowing in two directions at one time, if you just think in terms of Kirchhoff's current law ("The algebraic sum of currents entering and leaving a junction of conductors is zero."), it will be evident that it is not.

FULL-WAVE BRIDGE

Figure 2-19 is an example of the full-wave bridge combinational power supply. Part A shows a simplified schematic drawing, and part B shows the entire schematic, including the filtering network. It is a typical arrangement of the full-wave bridge combination supply, quite often called the "economy" power supply.

In part B of this figure, CR1 and CR3 form the full-wave rectifier circuit, and C2A, C2B, and L2 form the filter network. R1 is a current limiting resistor used to protect the diodes from surge

$$R(Z) = \frac{E}{I}$$

Fig. 2-17. This formula can be used to compare the values in the bleeder resistor voltage divider circuit.

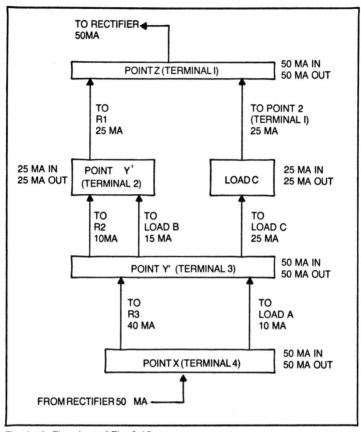

Fig. 2-18. Flowchart of Fig. 2-16.

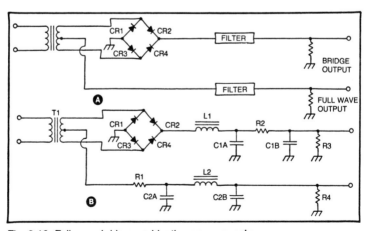

Fig. 2-19. Full-wave bridge combination power supply.

currents. R4 is the bleeder resistor and also assures that the power supply has a minimum load at all times. CR1 and CR4 form the bridge rectifier circuit with L1, R2, C1A, and C1B doing the filtering. R3 is a bleeder resistor and assures that the bridge always has a minimum load. Each circuit in itself works in the conventional manner. Troubleshooting will be the same as for the other power supplies discussed in this chapter entailing the no output and low output factors.

FULL-WAVE COMBINATIONAL

Figure 2-20 illustrates a full-wave combinational power supply with positive and negative outputs. It has one primary distinguishing feature compared to a bridge circuit: the center tapped transformer secondary. The components associated with the negative voltage output are CR1, CR3, L2, C2 and R2, while CR2, CR4, L1, C1, and R1 are the components in the positive voltage output. Transformer T1 is a component common to both supplies.

The operation of each full-wave rectifier is identical. When point A is negative with respect to ground, point B will be positive. This condition causes both CR1 and CR4 to conduct. Both these diodes are associated with different full-wave rectifiers. The negative power supply will be described first. This is the one associated with the conduction of CR1. Current through CR1 flows through L2, C2, and R2 and completes the path via ground to transformer centertap, and then to A. This is the path during the first half cycle while CR1 is conducting. At the same time this action is taking place, CR4 is conducting in the positive power supply. Since point B is positive and the lower half of the transformer secondary acts as a source, current from point B to ground flows up through C1, charging it as shown, L1, CR4, and back to the source. When the polarities at point A and point B are reversed (point A now being positive and point B being negative with respect to ground), the conducting diodes are now CR3 in the negative supply and CR2 in the positive supply.

Again, taking the negative supply first, CR3 conducts and the current path from point B and back to point B is through CR3, L2, up through C2 in the same manner as when CR1 was conducting, to ground and through the center tap of the transformer to point B. This completes the full cycle of operation for the negative supply, and CR1 and CR3 will be conducting alternately so long as there is an input.

At the time CR3 is conducting in the negative power supply,

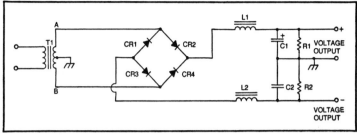

Fig. 2-20. Full-wave combination power with positive and negative output voltages.

CR2 is conducting in the positive power supply. Current from point A back to point A is going down the upper half of the transformer secondary to ground, up through C1 and L1, through CR2 and back to point A. Again, the charge path for C1 is the same as it was when CR4 was conducting.

In troubleshooting the combinational power supply, the problems are typically of the *no output* and *low output* type. You should realize that in the full-wave bridge combinational power supply, a low output can occur in either section, or the voltage of the full-wave rectifier can be normal while the bridge rectifier might indicate a low output. If the load currents are not excessive and the filter components have checked satisfactorily, the defective diodes in the first case would be assumed to be CR1 or CR3 and in the second case to be CR2 or CR4.

VOLTAGE MULTIPLIERS

Figure 2-21 depicts a simple half-wave rectifier circuit that is capable of delivering a voltage increase (more voltage output than voltage input), providing the current being drawn is low. It is shown that it is possible to get a larger voltage out of a simple half-wave rectifier so long as the current is low. If the current demand increases, the output voltage will decrease. This can best be explained by the use of RC time constant. The charge time for the circuit in this figure is very fast since the circuit elements in the capacitor's charge path are the diode, CR1, the surge resistor, R, and the secondary of the transformer. These elements combine to form a very low impedance, since CR1 is conducting during the charge time and the value of R is about 20 ohms. In comparison, the discharge path for the capacitor is through the load, which offers an impedance several hundred times higher than that of the charge path. The lower the load impedance, the greater the current. If the

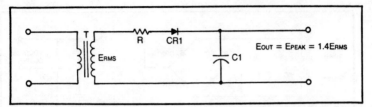

Fig. 2-21. Simple half-wave rectifier used to deliver an increased voltage output.

discharge path for the capacitor offers a lower impedance, the capacitor will discharge further, lowering the output voltage.

Rectifier circuits that can be used to double, triple, and quadruple the input voltage are very useful. All these circuits have one thing in common. They use the charges stored on capacitors to increase the output voltage. Figure 2-22 is a block diagram of a voltage multiplier circuit. The input is ac and the output is a dc multiple of the ac input.

HALF-WAVE VOLTAGE DOUBLER

The first voltage multiplier circuit is the half-wave voltage doubler. As the name implies, this circuit gives a dc output that is approximately twice that obtained from the equivalent half-wave rectifier circuit.

Figure 2-23 shows a typical half-wave voltage doubler circuit. While the circuit shown uses a transformer and the output voltage is positive with respect to ground, it could just as well operate as a negative voltage output by reversing the diodes. The transformer, which may be used to step up secondary voltage or as an isolation transformer, may also be eliminated with the proper choice of circuit elements.

When the top of the transformer secondary in this figure is negative, C1 will charge through conducting CR1 to approximately the peak of the secondary voltage. The direction of charge is indicated by the polarity signs. At this time, there is no output. On the next alternation, when the top of the transformer secondary is positive with respect to the bottom, C1 will discharge through the transformer and CR2, which is now conducting. C2, which is also in this discharge path, is charged to approximately twice the peak of the secondary voltage because the charge on C1 is in series with the applied ac and adds to the voltage applied to C2.

Since C2 receives only one charge for every cycle of operation, the ripple frequency, as in the half-wave rectifier, is the same as the input frequency. Also, as in the half-wave rectifier, C2 will dis-

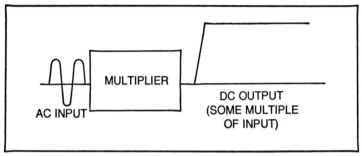

Fig. 2-22. Block diagram of a voltage multiplier circuit.

charge slightly between charging cycles so that filtering is required to smooth the output and give us a relatively pure dc.

The procedures for troubleshooting the half-wave voltage doubler are the same as those used for the half-wave rectifier. No output conditions might be caused by a defective transformer, defective rectifiers, an open C1, or short-circuited C2. The low output condition might be caused by a low input, rectifier aging, or excessive load current caused by a decrease in load impedance.

The voltage multiplier circuits which follow all have one thing in common with the circuit just described. They use the charge stored on a capacitor to increase the output voltage. As the voltage across C1 is added to the input voltage to approximately double the charge applied to C2, so will the charges on other capacitors add to the charge applied to an input capacitor to double, triple, or quadruple the output voltage.

FULL-WAVE VOLTAGE DOUBLER

The full-wave rectifier circuit can also be adapted to a voltage doubling circuit. Figure 2-24 depicts a basic full-wave voltage dou-

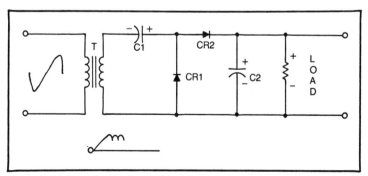

Fig. 2-23. Half-wave voltage doubler circuit.

33

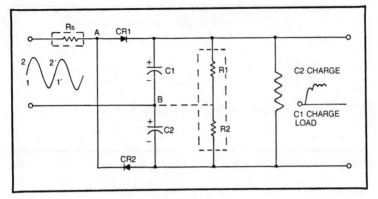

Fig. 2-24. Basic full-wave voltage doubler circuit.

bler circuit. Depending upon the circuit application, it may or may not use a power or isolation transformer. The resistor R_s is a surge resistor that is used to limit the charge current and protect the diode. It might not be necessary in some equipment and when used, it is placed in series with the ac source. Resistors R1 and R2 are not necessary for circuit operation but may be used to act as bleeder resistors to discharge their associated capacitors when the circuit is de-energized. When used, they also tend to equalize the voltages across C1 and C2.

The circuit operates much the same as the full-wave rectifier previously discussed, with the exception that now two capacitors are employed, each one charging to approximately the peak voltage of the input and adding their charges to provide an output. When point A is positive with respect to point B, C1 will charge through the conducting diode, CR1, and the source. It will charge to approximately the peak of the incoming voltage. On the next half cycle of the input, point A is now negative with respect to point B, and CR2 conducts, thus charging C2 in the direction indicated. The voltage across the load will be the total of the voltages across C1 and C2. C1 and C2 will be equal value capacitors, and R1 and R2 will also be equal value resistors. The value of R_s will be small, probably in the 20-500 ohm range.

VOLTAGE TRIPLER

Figure 2-25 depicts a typical voltage tripler circuit with waveforms and circuit operation. Part A shows the complete circuit. In Part B, C1 is shown charging as CR2 is conducting. In Part C, C3 is illustrated charging as CR3 is conducting. Part D reveals the

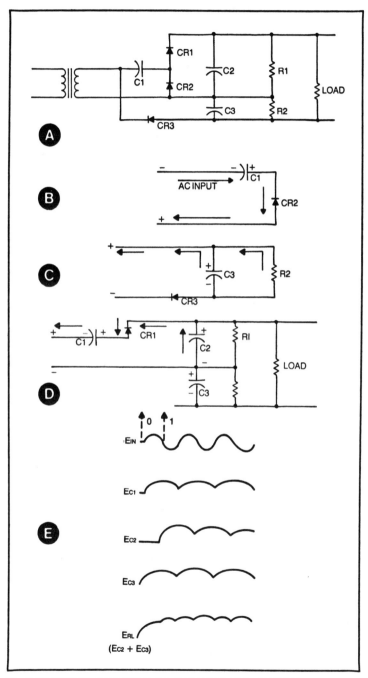

Fig. 2-25. Typical voltage tripler circuit operation and waveforms.

charge path for C2 while CR1 is conducting. In part E, a comparison is made of the input signal and its effects on the voltages felt across C1, C2, C3, and the load. The following explanation uses this figure as the operating device.

Close inspection of Part A should reveal that removal of CR3, C3, R2, and the load resistor results in the voltage doubler circuit previously described. The connection of circuit elements CR3 and the parallel network of C3 and R2 to the basic doubler circuit is arranged so that they are in series across the load. The combination provides approximately three times more voltage in the output than is felt across the input. Fundamentally, then, this circuit is a combination of a half-wave voltage doubler and a half-wave rectifier circuit arranged so that the output voltage of one circuit is in series with the output voltage of the other.

Part B shows how C1 is initially charged. Assume the input is such that CR2 is conducting. A path for charging current is from the right hand plate of C1 through CR2 and the secondary winding of the transformer to the left hand plate of C1. The direction of current is indicated by the arrows.

At the same time that the above action is taking place, CR3 is also forward biased and is conducting, and C3 is charged with the polarities indicated in Part C. The arrows indicate the direction of current. There are now two energized capacitors, each charged to approximately the peak value of the input voltage.

On the next half cycle of the input, the polarities change so that CR1 is now the conducting diode. Part D indicates how capacitor C1, now in series with the applied voltage, adds it's potential to the applied voltage. Capacitor C2 charges to approximately twice the peak value of the incoming voltage. As can be seen, C2 and C3 are in series and the load resistor is in parallel with this combination. The output voltage, then, will be the total voltage felt across C2 and C3, or approximately three times the peak voltage of the input.

Part E indicates the action taking place using time and the incoming voltage. At time zero (t_0), the ac input is starting on its positive excursion. At this time, the voltage on C1 and C3 is increasing. When the input starts to go into its negative excursion, t_1, both C1 and C3 start to discharge and the voltage across C2 is increasing. The discharge of C1 adds to the source voltage when charging C2 so that the value of E_{C2} is approximately twice the value of the peak value of the input. Since E_{C2} and E_{C3} are in series across the load resistor, the output is their sum.

The charge paths for capacitors C1, C2, and C3 are comparatively low impedance when compared to their discharge paths.

Therefore, even though there is some ripple voltage variation in the output voltage, the output voltage will be approximately three times the value of the input voltage. The ripple frequency of the output, since capacitors C2 and C3 charge on alternate half cycles of the input, is twice that of the input ripple frequency.

Troubleshooting the voltage tripler circuit follows the general practice given for rectifier circuits. The two general categories of failure are no output or low output. For the no output condition, look for a no input condition, a lack of applied ac, a defective transformer, or a shorted load circuit. For a low output condition, check the input voltage. Low output voltage might also be a result of any of the following: rectifier aging causing an increased forward resistance or a decreased reverse resistance; any leakage in the capacitors or decrease in their effective capacitance; or an increase in the load current (decrease in load impedance).

VOLTAGE QUADRUPLER

Figure 2-26 shows a typical voltage quadrupler circuit. Essentially, this circuit is two half-wave voltage doubler circuits connected back-to-back and sharing a common ac input. In order to show how the voltage quadrupler works, Fig. 2-26 has been shown as two voltage doublers. The counterparts of the one circuit are shown as a prime (') in the second circuit. (C1 in the first circuit is the same as C1' in the other circuit; CR1 and CR1' conduct at the same time, etc.) When the circuit is first turned on, it will be assumed that the top of the secondary winding, point A, is negative with respect to B. At this instance, CR1 is forward biased and conducting, allowing C1 to charge. On the next alternation, point B is negative with respect to point A. At this time, two things are going to occur: (1) C1, which was charged to approximately the peak voltage across the secondary winding, will aid the source, and since C2 is now conducting, C2 will be charged to approximately twice the

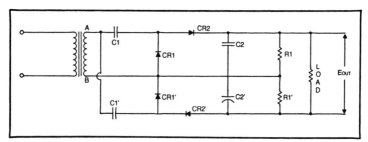

Fig. 2-26. Typical voltage quadrupler circuit.

incoming voltage; and (2) CR1' will conduct charging C1' to approximately the peak voltage of the input. During the following alternation, C1' adds to the input which allows CR2' to conduct, charging C2' to twice the input voltage.

When CR2 conducts, C1 will aid the input voltage in charging C2 to approximately twice the peak voltage of the secondary; and CR2' conducting at the same time, charges C1'. On the next alternation, CR1' conducts; and since C1 is in series with the input, it aids in charging C2' to twice the peak of the secondary. The voltages across C2 and C2' add to provide four times the peak secondary voltage in the output.

SUMMARY

The basic dc power supply is quite simple as far as electronic circuits go. Using the material contained in this chapter, simple power supplies may be quickly and easily constructed. However, most modern solid-state circuits which require operating current from a dc power supply cannot directly use the outputs from the basic supplies discussed. A later chapter will deal specifically with voltage regulation, which is usually required by many solid-state circuits.

You've learned that the power supply transformer and the rectifier configuration will directly impact the voltage at the power supply output. This is also affected by the type of filter arrangement used. A single type of transformer, then, could be used to deliver several different voltage values at the output of a dc power supply, depending upon the rectifier circuit and the filter. This principle is used in power supply design to allow one transformer (which is often the most expensive part of a power supply circuit) to serve in many different dc power supply applications.

Chapter 3

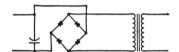

Dc Power Supply Components

Before continuing further, you must understand the various components that are used to build power supplies. These are used in later construction projects presented. Fortunately, most dc power supplies utilize easily available electronic components which are generally low in cost and may be purchased from most electronic hobby stores. Occasionally, a special component will be needed which may require you to do a little searching, but this is the exception rather than the rule.

TRANSFORMERS

The transformer is considered to be the heart of the dc power supply and is often the most expensive component. Surprisingly enough, transformers are ac devices and will not operate directly from direct current. The output from a transformer is alternating current and must be rectified and filtered for a pure direct current output.

A transformer is basically two coils of wire which share a single core. Sometimes two cores are used, one for each winding, but the coils are physically positioned so that energy is transferred from one to the other.

The main job of the transformer is to do just what its name implies: it transforms voltage of one value to another. A transformer consists of a primary winding and a secondary winding. This is shown schematically in Fig. 3-1. The winding on the left is the

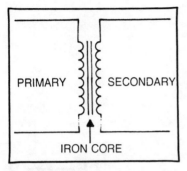

Fig. 3-1. The basic transformer consists of a primary winding and a secondary winding sharing a common iron core.

primary, the core is in the center, and the secondary winding is seen on the far right. In most dc power supply applications, the primary winding receives current from the 115 volt household line. The primary winding is built to receive this value of voltage (at a frequency of 60 Hertz in most cases) and transform it to a higher or lower voltage value at the output of the secondary winding. In some cases, the output at the secondary winding will be the same as that at the primary. These types of transformers are used for isolation purposes and are called isolation transformers. An isolation transformer is used to prevent some types of equipment from picking up noise and other interference from the ac line. Due to the nature of the transformer, there is no direct or dc connection between the primary and secondary.

In most cases, however, the output of the secondary winding will be different from that of the primary. While transformers are available with many different secondary output values, many are standardized in that values of 6, 12, 20, and 40 volts are often seen. Other transformers may have secondary output values of 2.5, 5, 7.5, and 25 Vac and are also quite common. The devices just mentioned are referred to as low-voltage transformers, in that they transform the 115 volt line supply to a lower value.

Medium-voltage transformers will usually have outputs from 115 to around 400 volts, and high-voltage transformers will produce secondary outputs of around 500 to several thousand Vac. One will often hear the term step-up or step-down used when describing transformer functions. The low-voltage transformers discussed are a good example of step-down transformers, in that the input voltage to the primary winding is stepped down to a lower value at the secondary. The medium- and high-voltage devices step the input voltage up to a higher value.

Figure 3-2 shows a pictorial diagram of a typical iron core transformer. The core is actually that portion of the metal structure

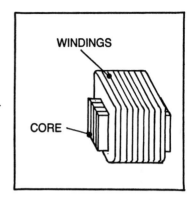

Fig. 3-2. A typical iron core transformer.

which lies inside the windings. The outer portion is called the frame. Transformer cores are usually composed of laminated steel, which is assembled in strips. Figure 3-3 shows these strips. Each is coated with an insulating lacquer and then placed one on top of the other. The iron core serves to magnify the inductance and coupling of the primary and secondary windings. It is possible to build a transformer without an iron core; but at frequencies of 60 Hz, the overall size of the device would be enormous. When a transformer is composed of windings mounted on a magnetic material, the overall value of inductance is multiplied by a factor which is equivalent to the core size. Therefore, transformers with large cores will have higher power ratings than those with smaller cores. The size of the

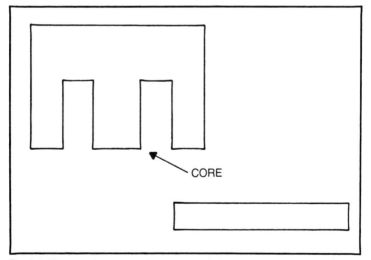

Fig. 3-3. The transformer core is constructed of layers of laminated steel strips which are coated with an insulating lacquer.

core is partially dependent upon the ac line frequency. Some specialized power transformers are designed to operate at a frequency of 400 Hz rather than the standard 60 Hertz. As frequency increases, the size of the core will decrease, assuming both devices are rated at the same power output. In other words, as far as transformers are concerned, high frequency operation is more efficient than low frequency operation. The reader should be warned at this point that 400 Hertz transformers are available from many war surplus outlets at dirt cheap prices. These will not work from the 60 cycle line and no easy conversion is possible. Likewise, it is not practical to step up the frequency of the 60 Hertz ac line.

While the iron core enables 60 Hz transformers to be physically small in size when compared with coreless construction, it does introduce certain problems. When an iron core transformer is operational, small currents will flow within the core itself. These are called *eddy currents*, and they set a limit on the amount of current that can be supplied by the transformer secondary. Other factors are also included in the total power output rating. These include wire size of the primary and secondary windings. A single conductor will only carry so much current before it becomes overheated. When this occurs, tremendous losses are incurred and the heating effects may cause the conductor to break or the transformer insulation to melt and be destroyed.

Construction of power supply transformers involves designs which attempt to keep the magnetic path around the core as short as possible. Shorter magnetic paths dictate fewer turns in the primary and secondary windings for a specific voltage input and output. Since every conductor contains a specific amount of resistance, fewer turns mean shorter lengths and thus less internal resistance. When internal resistance is kept to a minimum, the transformer will operate most efficiently due to decreased heat (I^2R) losses.

TRANSFORMER POWER RATINGS

The amount of power that a transformer will deliver is dependent upon several factors. These include conductor size of the primary and secondary windings, the size of the core, and also the type of filter and rectifier circuits used. The power rating of most transformers is given in *watts* or in *volt-amperes*. Volt-ampere ratings are most common and simply indicate the product of the secondary output voltage times the rated current. If a transformer has a 100-volt secondary value at 1 ampere, then the volt-ampere rating would be around 100 watts. Actually, the rating would be a bit higher than this because more than 100 watts of power would be drawn from the ac line by the primary winding. The difference here is the losses within the transformer itself. Normally, these are very low and most modern transformers operate with efficiencies in excess of 90 percent.

You learned earlier that the type of filter and rectifier circuits used with the transformer will also be a determining factor on power which may be delivered by the transformer. A capacitor input filter delivers high peak values of voltage and current and causes a higher thermal effect in the secondary transformer winding. The average values of voltage and current obtained with a choke input filter do not stress the secondary as much and result in cooler operation. Capacitor input filters often necessitate a higher transformer power rating because the power delivered to the load is often less than that demanded from the transformer.

Some power supply transformers will have one primary winding and one secondary winding. This means that the input voltage and the output voltage are fixed. Others will have a single primary winding and multiple secondaries. This latter design is shown in Fig. 3-4. Here, the secondary winding consists of three different coils, one of which will supply a moderate voltage, the other a slightly higher or lower voltage, and the third a low voltage. This allows one transformer to be used with three separate rectifier and filter circuits to effectively deliver three power supplies from a single device.

Another transformer arrangement which is commonly seen has a single primary and secondary winding, but the secondary has been *tapped* at two or more points. These taps are simply connections which are made to the secondary winding and brought out of the frame for electrical contacts with other circuits. This type of transformer is shown schematically in Fig. 3-5. Note that the transformer

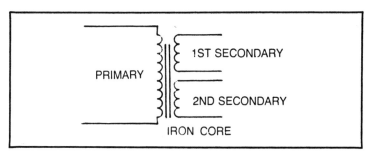

Fig. 3-4. Schematic representation of a transformer having multiple secondary windings.

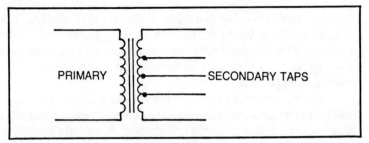

Fig. 3-5. Schematic of a transformer having a tapped secondary winding.

has only one secondary winding, but several portions of this winding are made accessible.

Figure 3-6 shows a very common type of transformer which is called a center tapped device. Here, the secondary winding contains one tap at the center of the coil. If the voltage rating of the entire secondary winding is 100 Vac, then the rating between the center tap and either of the remaining leads will be one half this value, or 50 Vac. This type of transformer is required for a full-wave center tap rectifier configuration. This was discussed earlier and using the transformer under discussion would result in a dc output of approximately 50 Vdc under moderate to heavy loading. This same transformer could be used with a full-wave bridge rectifier circuit by simply disregarding the center tap and using the two outside secondary leads. This circuit would deliver twice the former output, or 100 Vdc under moderate to heavy loading.

It would seem that we would get twice the output by using the bridge rectifier circuit. This is absolutely correct if, by output, we are referring to voltage alone. However, remember that the transformer also has a power rating which is directly dependent upon voltage and current. If this transformer were rated at 50 watts, then we could draw 1 ampere of current at 50 volts when using the full-wave center tapped circuit (1 ampere × 50 volts). Using the full-wave bridge rectifier, we could still draw *only* 50 watts from the transformer. This would mean that at a 100 Vdc output, only one-half ampere of current could be demanded. 0.5 amperes times 100 volts still equals 50 watts. If we attempted to draw 1 ampere of current at this higher voltage, a total power consumption of 100 watts would be demanded. This is 100% more than the transformer was designed to deliver.

In certain applications, it is possible to draw more current and thus power from a transformer than it is apparently designed to deliver. Transformers often carry two different types of ratings.

One may be labeled as 50 volt-amperes CCS, while another may say 50 volt-amperes ICAS. The former rating indictes that 50 watts of power may be drawn from the transformer in *continuous commercial service (CCS)*. Continuous commercial service means just about what it says. This transformer will deliver 50 watts of power around the clock in continuous service. An *ICAS* rating stands for *intermittent commercial and amateur service*. The latter transformer is designed to deliver 50 watts of power for a limited period of time with an equal amount of time of non-use to allow for cooling. It is often difficult to figure the rating of one type of service when the other is placarded on the transformer. For example, a 50 watt CCS transformer might safely deliver 100 watts of power when used in ICAS applications. I usually figure an increase of approximately 50% when using a CCS device for ICAS service. This would mean that the 50 watt CCS transformer would most likely deliver 75 watts of output power for moderate lengths of time without exceeding its rating.

As was previously mentioned, there is an abundance of transformers available through war surplus channels. While some of these may be designed to operate from unusual primary voltage values and frequencies, most are perfectly applicable to 115-230 volt, 60 Hz lines. In most cases, transformers designed for military applications are given very conservative ratings. It is often possible to draw twice the rated power output from these devices without any undue heating or damage. In some cases, I've drawn triple the power from some transformers without any harmful effects occurring. Different transformers will exhibit different power capabilities, and it is usually necessary to experiment before you really know just what can be had from a specific unit. Generally speaking, military transformers which are designed to deliver an output current of 1 ampere, for instance, in a full-wave center tapped configuration will safely deliver the same 1 ampere output when

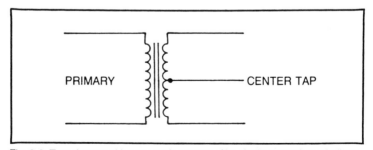

Fig. 3-6. Transformer with a tap at the center of its single secondary windings.

connected to a full-wave bridge rectifier at twice the voltage. This effectively means that you are drawing twice the rated current output. Again, experimentation is required and checks should be made for voltage drop at the secondary winding when drawing higher than rated amounts of power. It is also important that the body temperature of the transformer case be watched carefully during these initial stages.

TAPPED PRIMARY WINDINGS

Earlier, you learned about multiple and tapped secondary windings. There are also multiple and tapped transformer primaries. A popular arrangement is shown schematically in Fig. 3-7. Here, the transformer contains two primary windings, each rated at 115 volts ac. This transformer may be operated at full power from 115- or 230-Vac. For 115-volt operation, the two windings are connected in parallel, as is shown in Fig. 3-8. For 230-volt operation, the windings are connected in series, as shown in Fig. 3-9. Two leads from each winding are brought to the outside of the transformer case, so it is only necessary to connect these external leads as desired.

In some instances, it is desirable to power a transformer from the higher 230 Vac than at the lower voltage. This is especially true in high power circuits where large amounts of current must be drawn from the primary ac line. For example, if a transformer is connected to a circuit which has a power drain of 1,000 watts, at 115 Vac, nearly 10 amperes will be drawn from the ac line. But at 230 Vac, only half this amount of current is drawn to deliver the same power input to the transformer. (115 V × 10 amperes = 1,150 watts; 230 V × 5 amperes = 1,150 watts.) The electrical wiring in your home is rated to carry only so much current. Here, the current rating is all that matters; the voltage really has no effect on losses. If

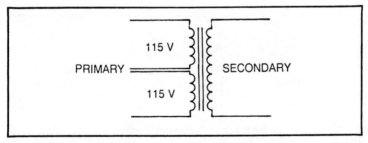

Fig. 3-7. Schematic representation of a transformer with two primary windings, each rated at 115 Vac.

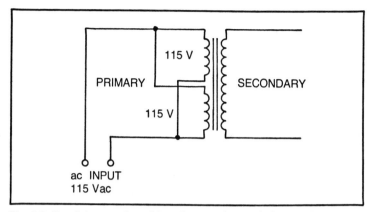

Fig. 3-8. Parallel connection of transformer primary windings yields 115-volt operation.

a specific power loss exists in the house wiring at a drain of 10 amperes, this same loss will occur regardless of the line voltage. So, at 230 volts, we are incurring less heating losses in the household wiring than at 115 volts, even though the same amount of power is being drawn. High current demands on the ac line are indicated by household lights dimming or flickering as power is being demanded from a transformer.

Figure 3-10 shows another type of transformer primary arrangement which uses a single winding that has many taps. Household current in the United States is typically delivered at a voltage

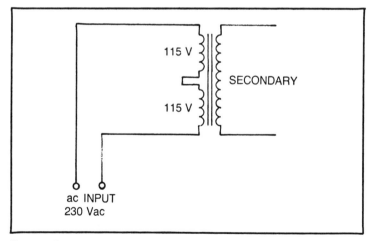

Fig. 3-9. Series connection of two 115-volt primary windings yields 230-volt primary operation.

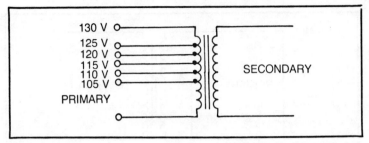

Fig. 3-10. Power transformer with tapped primary winding for operation from 105 to 130 Vac.

of about 115 Vac. This will vary from location to location, however, and at certain times of the year. In the summer, for instance, air conditioners operating around the clock can lower the voltage delivered by your electric company. Air conditioners are high current devices, and many thousands of them can have substantial effects on voltage over the entire service area of your power utility. In this case, your voltage in the summer could drop to as low as 105 Vac. In other areas, voltage levels may run high and be measured at 125 Vac or even 130 Vac. In most applications, this does not make a great deal of difference, but in others, wide voltage fluctuations can completely disrupt operation.

The previous figure shows a transformer which can help to overcome these problems. The primary winding is tapped so that different values of input voltage can be used to deliver the correct secondary voltage. If the input voltage is 125 Vac, then the appropriate taps are used. During the cooling season, the 105 volt tap might be used if this is the average level to which the household voltage drops. In both cases, the secondary output voltage will remain virtually unchanged. This kind of arrangement is often used on high-voltage transformers designed for commercial broadcasting services. Here, high amounts of power are required from the transformer and voltage drops can reduce overall transmitting efficiency. One must be careful when setting these taps. For example, if the 105 Vac taps are used and the voltage suddenly increases to 130 Vac, then the secondary winding will produce an output voltage which is nearly 30 percent higher than normal.

COMBINING TRANSFORMERS

It has already been stated that transformers are ac devices and as such, the primary and secondary windings are interchangeable to some degree. These are double-ended components, in that the

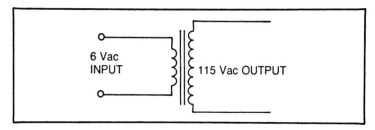

Fig. 3-11. A transformer is a reversible device. Here, a filament transformer which is usually operated with a 115-volt primary source is used with a 6-volt input at its former secondary winding to produce a 115 Vac output.

secondary winding may also serve as the primary. The primary winding may also be reversed. By this we mean that a transformer with a 115 volt primary and 6 volt secondary can be used in one of two ways. The standard arrangement would be to connect a 115 Vac source at the primary winding for an output of 6 volts at the secondary. Alternately, a 6 volt ac input may be applied at the secondary winding for an output of 115 Vac at the primary. In this latter role, the former secondary winding becomes the primary and the former primary is now the secondary. This is shown in Fig. 3-11. This feature can come in handy when the need arises to obtain a transformer output voltage which is not available from any one unit you happen to have on hand. For example, let's assume that you need a transformer which has a 115 Vac input and an output of 230 Vac. Assume further that all you have on hand is two low-voltage transformers, one with 115 Vac input and the other with a 230 Vac input at their primaries. Both transformers have secondary windings rated at 12 Vac. Figure 3-12 shows that the 115 volt primary of one transformer is connected directly to the house line, while the 12 volt secondary is connected to the secondary of the transformer with the 230 volt primary. The end effect is a single transformer

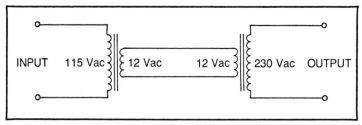

Fig. 3-12. A combination of two transformers, both with 12-volt secondaries and one with a 115-volt primary while the other is powered from 230 Vac. This arrangement yields a single complex transformer with 115-volt primary input and 230-volt secondary output.

with a 115 Vac primary and a 230 volt secondary. The first transformer steps down the 115 volt house current to 12 Vac. This serves as a primary drive to the 12 volt winding of the second transformer, which steps up this voltage to 230 Vac. This is just one example of what can be had by combining transformer primaries and secondaries. The total power rating of the final secondary output will be equal to the lowest volt-ampere rating of either transformer. Let's assume that both transformers have a 12 watt rating, delivering a 12-volt secondary output at 1 ampere. Since the second transformer has been reversed, this means that the maximum input power to the second 12 volt winding cannot exceed 1 ampere. This also means that the maximum amount of current which can be drawn from the final 230 volt winding will be 50 milliamperes, or 0.05 amp. (230 V × 0.05 ampere = 11.5 watts)

SERIES AND PARALLEL CONNECTION OF TRANSFORMERS

Transformers may also be combined in a different manner. This involves the connection of primary windings in parallel or series. The same can be done with the secondary windings. Let's assume you need a transformer with a 230 volt primary and a 6 volt secondary. If you have two 6 volt transformers with 115 volt primaries, what you need can be built in a few minutes. An earlier portion of this chapter discussed transformers with two primary windings. This is basically what you have here but in two separate components. This arrangement works best if both transformers are nearly identical. Figure 3-13 shows that the two primary windings are connected in series. Their secondary windings are connected in parallel. Now, you effectively have a single transformer with a 230 volt primary and a 6 volt secondary. The total power output will be double that of a single transformer, assuming that both are nearly identical. Due to the makeup of the windings, it is necessary to connect the primary leads into what is known as a *series aiding* circuit. Transformer windings may be wound left to right to left. Either way, a single unit will operate properly. But, should you connect the transformers discussed above with one coil going left to right and the other right to left, the circuit will not work. If two identical units are used, you simply connect the right hand lead of one transformer to the left lead of the other, as in Fig. 3-14A. Two transformers made by different manufacturers may be wound differently and could require that the right hand lead of one be connected to the left hand lead of the other, as shown in B. This is usually not a great problem and should not result in transformer

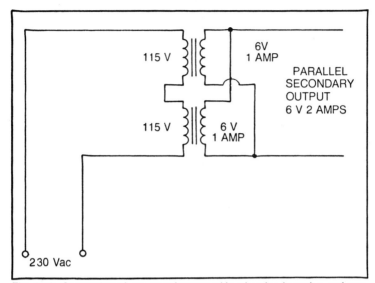

Fig. 3-13. Connection of two transformers with primaries in series and secondaries in parallel.

damage should you accidentally reverse the connections. It's a good idea to connect the primary windings as you think they should be and then measure the output voltage from *one* of the secondary windings with an ac voltmeter while primary power is applied. If you get no output, simply transpose the primary windings.

The same is true of the secondary windings, which in this case are wired in parallel. If both transformers are identical, the left leads of both transformer secondaries are connected together, as are the right leads. Once you have established that the primary windings are connected correctly, make your best guess at the parallel connections of the secondary windings and then measure the output voltage at this latter point. If you get nothing, transpose the secondary connections and everything should be in order.

This arrangement is very practical and it is quite easy to double the output voltage by reconnecting the secondary windings in series. This is shown in Fig. 3-15. Now, the overall transformer has a 230 volt primary and a 12-volt secondary. If you wanted to switch back to 115 volt primary operation, all that's necessary is to connect the primary windings in parallel and you end up with a transformer that operates from 115 volts ac and produces a 12 volt output when the secondaries are wired in series.

This leads to another arrangement where you may need more power output than can be obtained from one transformer. Figure

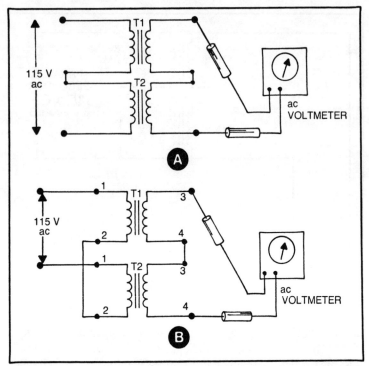

Fig. 3-14. If the primary connections in A result in zero output, the connection in B should work.

3-16 shows two transformers, each with 115 volt primary and a 12 volt secondary and rated at 1 ampere. By combining the two in parallel, you end up with an overall transformer that has the same primary input voltage and secondary output voltage but at a rating of 2 amperes. You can even combine three or more transformers for higher power ratings, although size may become a factor. Should you connect the secondary windings in series, your output voltage would double, as would your output power; but you still must draw the same amount of current from this complex component as you would from a single transformer. In other words, if each transformer is rated at 1 ampere output from a 6 volt secondary, then when the secondaries are wired in series, you can still draw only 1 ampere. Your power is doubled because your voltage is twice that of one unit. On the other hand, if you connected the two secondary windings in parallel, secondary output voltage would still be 6 Vac but you could draw twice the current. Either way, you end up with a total combined output of 12 watts.

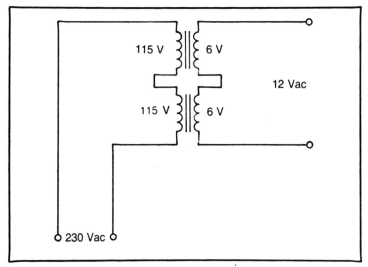

Fig. 3-15. Wiring of two identical transformers with secondaries in series, as are primary windings.

LOW VOLTAGE OPERATION

If your transformer is designed to operate at a primary input voltage of 115 Vac, the secondary output is solely dependent upon this input value. It was previously stated that drops in power line voltage can play havoc with transformer outputs. If your primary

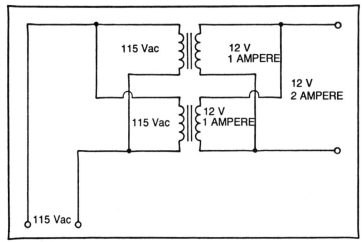

Fig. 3-16. Parallel connection of identical transformers yields 115-volt primary operation and a 12-volt secondary. This system is rated at twice the power output as with one unit, as the secondary current drain may be doubled.

voltage decreases by 10%, then the secondary voltage will do the same. Upping the input voltage will have the same percentage effect on the secondary output. This can be used to good advantage in some limited applications. For example, a transformer which produces a secondary output of 12 Vac with a primary input of 230 Vac will produce half this amount of output when driven with 115 Vac at the primary. A simple dc power supply with no electronic regulation which produces 12 Vdc at its output when the transformer is driven by a 230 volt line will produce a corresponding 6 Vdc output if the primary voltage is dropped to 115 volts. If you have access to both a 230 volt and 115 volt line outlet, then a single power supply can be used to deliver two different dc outputs. *Caution*: The reverse of this will not work. While you can power a transformer with a 230 volt primary at a value of 115 volts ac, you cannot apply the higher voltage to a transformer designed to operate with a primary value of 115 volts. One power supply project in a later chapter takes advantage of a transformer with a 230 volt primary to produce a dual dc output potential.

All of the transformers discussed in this book are designed to be powered from standard household current rated at 115-230 volts ac single-phase. Multiphase transformers are used for industrial and high-power applications and require special wiring from the power company. These types of applications are not within the scope of this text.

SOLID-STATE RECTIFIERS

A rectifier is a device that will conduct current in one direction but not in the other. These devices and circuits have been discussed in an earlier chapter, but a bit more should be said about their uses in power supply circuits. Solid-state rectifiers have almost completely replaced the old thermionic tube type devices of a few decades ago and are far more efficient to work with. Whereas the tube type rectifiers required power for their filaments, solid-state designs do not. The latter are very thin and require very simple mounting considerations.

The main ratings that we are concerned with when using rectifiers for power supply design in PRV and I_o. The former is the *peak reverse voltage* rating, while I_o indicates the maximum amount of current the device is designed to conduct. The forward current (I_o) rating is quite easy to arrive at. Simply figure the maximum amount of power your supply must deliver and add a bit as a safety measure. For example, if a power supply is to deliver 600 milliamperes of

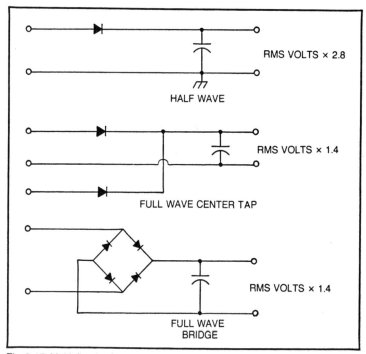

Fig. 3-17. Multiplication factors used to determine PRV ratings for the three major rectification circuits.

current, diodes should be chosen with a 1 ampere (1,000 milliamperes) I_o rating.

The peak reverse voltage rating will be determined by the transformer voltage and by the type of rectifier circuit in which each diode is used. Figure 3-17 shows the multiplication factors used to determine PRV ratings for the three major rectifier circuits. A half-wave rectifier will require a diode with a rating which is equal to 2.8 times the transformer secondary voltage. It's easier to figure 3 times this value in most practical applications. This also adds a bit of a safety margin. For example, if your half-wave circuit is to be connected to a transformer with a secondary voltage value of 12 Vac, then the minimum PRV rating should be 36 volts. Power supply rectifiers are usually available with PRV ratings starting at 50 volts, so a 50 volt component would be more than adequate.

A full-wave center tapped circuit will also require diodes rated at 2.8 times the secondary voltage. This means something different than with the former half-wave circuit. Due to full-wave center tapped design, the multiplication factor is applied to one-half of the total secondary voltage, since only one-half of the transformer winding is used on any one cycle of rectification. If your transformer is rated at a total secondary voltage of 12 Vac center tapped, then you would multiply 2.8 times one-half of 12 volts, or 6 volts. It's usually simpler just to use a multiplication factor of 1.4 times the total secondary voltage.

The full-wave bridge circuit requires diodes rated at 1.4 times the total secondary voltage. This circuit is not as complicated to figure as is the center tapped arrangement.

Assuming that all of these supplies use transformers with 12 volt secondaries, the half-wave design would require a PRV rating of about 36 volts, the full-wave center tapped circuit approximately 17 volts, and the bridge design 17 volts. You can see that any of these circuits could be safely constructed using rectifiers with 50 volt PRV ratings.

Like transformers which you learned about earlier, diodes may be combined in series and in parallel to increase ratings by a multiplication factor equivalent to the number of total components. PRV ratings are increased by connecting diodes in series. This is shown in Fig. 3-18. Here, three 50 PRV diodes are connected in a *series string*. The total PRV rating of the combined circuit is now three times 50 PRV, or 150 PRV. The current rating will stay the same. If each diode is rated at 1 ampere at 50 PRV, then the string will be rated at 1 ampere but at 150 PRV.

Rectifiers are often connected in series strings, especially in moderate to high voltage applications. Components with PRV ratings of 1,000 volts are very common and inexpensive. However, when this 1,000 volt level is exceeded, individual diodes become more and more expensive. A 3,000 volt diode may cost in excess of ten dollars, but three 1,000 volt units can probably be had for less than fifty cents through industrial surplus outlets. It's far more practical in many applications to simply series-connect several diodes to arrive at the desired PRV rating. Diodes are available in many different current ratings starting at about a half of one ampere to several hundred or even a thousand amperes. It is rarely necessary to connect a number of these units for higher current ratings,

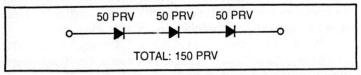

Fig. 3-18. Series connection of diodes increases the PRV ratings of diode rectifiers.

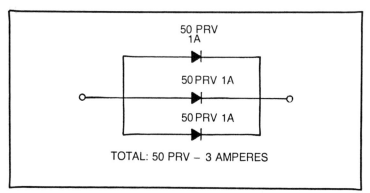

Fig. 3-19. Parallel connection of diode rectifiers.

except in a few instances. When two or more diodes are connected in parallel, their forward current ratings multiply. Figure 3-19 shows the previous three 50 PRV, 1-ampere diodes connected in a parallel configuration. The combined rating of this circuit is 50 PRV at 3 amperes. Notice that the forward current rating has been multiplied by the number of diodes in the circuit, while the reverse current rating remains the same. As many diodes as desired may be connected in parallel and the forward current rating will continue to increase by the number of total components.

While seldom used in modern design, it is possible to combine rectifiers in series-parallel arrangements. Figure 3-20 shows such a combination, which uses nine 50 PRV, 1-ampere components in series-parallel. Basically, the circuit shown consists of three series strings of three rectifiers wired in parallel. The total rating of this circuit will be 150 PRV at 3 amperes. Obviously, the cost of the nine

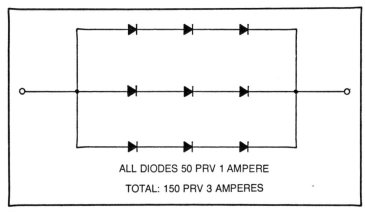

Fig. 3-20. Series-parallel connection of nine diode rectifiers.

components needed for this circuit would be far more than the price of a single diode which might typically have a rating of 200 PRV at 5 amperes. Such units are available through surplus markets for less than one dollar. True, you might be able to purchase individual 50 PRV, 1 ampere diodes through the same surplus channels for five cents each, but the wiring of nine units into a single circuit is far more complex than simply inserting one. Even here, the cost would still be higher, because when solid-state rectifiers are combined in series, parallel, or series-parallel, additional protective components are needed.

PROTECTIVE COMPONENTS

When wiring solid-state rectifiers in series, there is a good possibility of mismatch between components in the series string. Solid-state rectifiers used for power supply applications have a certain amount of internal resistance. Each will drop approximately 0.3 volt at the dc output. While internal resistance from component to component (assuming all are made by the same manufacturer and are rated equally) do not vary greatly, there is some difference and it is possible for a single rectifier in the string to take nearly all of the brunt of the peak reverse voltage. To overcome this, series strings of rectifiers are usually protected by wiring a high value, low wattage carbon resistor in parallel with each diode. This is shown in Fig. 3-21. Typical resistor values are in the neighborhood of 470,000 ohms (470 kohms). These parallel resistors are incorporated to better align the internal resistance of each rectifier. This provides a better match and each component handles an equal share of the overall load.

When power supplies are switched on and off, voltage spikes can occur. These are of very short duration, but can instantly destroy one or more rectifiers in the string. To overcome this problem, ceramic disk capacitors are usually wired in parallel with each rectifier. The completed circuit, with all protective elements

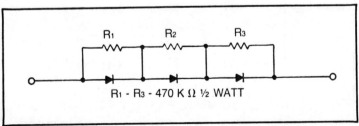

Fig. 3-21. Equalization resistors assure diode protection in series strings.

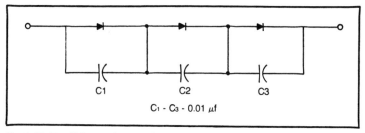

Fig. 3-22. Parallel ceramic capacitors protect series diodes from voltage spikes.

attached, is shown in Fig. 3-22. Each capacitor will have a typical value of 0.01 microfarad at 1,000 volts dc.

The cost factor of series strings of rectifiers is now becoming clearer. The three rectifiers must be wired in a circuit which also includes three resistors and three capacitors. Usually, the protective components are relatively low in price and the overall circuit costs will be less in high-voltage applications than if a single diode rated at 3 kilovolts or more were used. The same is not true of some parallel connections.

The parallel connection of diodes does not usually involve the use of parallel connected resistors and capacitors. But due to internal resistance differences, series resistors are often required to prevent one diode from passing most of the current output from the supply. If the resistors are omitted from the circuit, one diode which has a little less internal resistance may draw most of the current and be destroyed in a brief period of time. The series resistors are typically of a very low ohmic value and are chosen to deliver a voltage drop of about 1 volt at peak current. In a 1-ampere supply, this dictates a resistor value of 1 ohm. Low value resistors can cost more than several rectifiers in some instances. Figure 3-23 shows a typical protective circuit.

The cost factor of protective circuitry goes even higher when series-parallel strings of rectifiers are used for power supplies. Here, series and parallel resistors are required in addition to parallel capacitors. Figure 3-24 shows a typical circuit configuration.

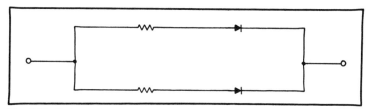

Fig. 3-23. Series resistors equalize parallel-connected rectifiers.

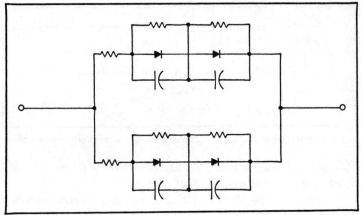

Fig. 3-24. Series-parallel rectifier circuit with all protective components.

It may be impossible to avoid the series connection of 1,000 PRV rectifiers in circuits designed for high voltage applications. While series connections are more complex, this may be the cheapest way to go for high voltage power supply design. Parallel connection of rectifiers is not nearly as attractive and due to the low price of high current components, one rarely has to resort to the complexities and extra expense involved in these types of connections.

CAPACITANCE

Capacitance is defined as the property of an electrical device or circuit that tends to oppose a change in voltage. Capacitance is also a measure of the ability of two conducting surfaces, separated by some form of nonconductor, to store an electric charge. The device used in electrical circuits to store a charge by virtue of an electronics field is called a capacitor. (The larger the capacitor, the larger the charge that can be stored.)

The simplest type of capacitor consists of two metal plates separated by air. A free electron which is inserted in an electrostatic field will move. The same is true, with qualifications, if the electron is in a bound state. The material between the two charged surfaces of Fig. 3-25 (air in this case) is composed of atoms containing bound orbital electrons. Since the electrons are bound, they cannot travel to the positively charged surface. Therefore, the resultant effect will be a distorting of the electron orbits. The bound electrons will be attracted toward the positive surface and repelled from the negative surface. This effect is illustrated in Fig. 3-26. In A of this

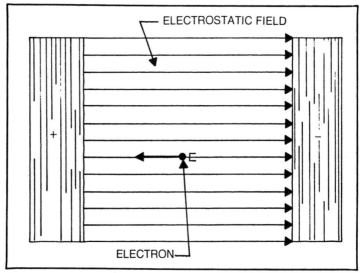

Fig. 3-25. Electron movement in an electric field.

figure, there is no difference in charge placed across the plates, and the structure of the atom's orbits is undisturbed. If there is a difference in charge across the plates, as shown in part B, the orbits will be elongated in the direction of the positive charge.

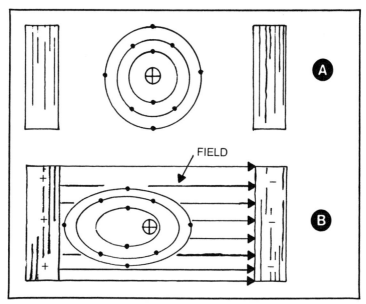

Fig. 3-26. Electron orbits with and without the presence of an electric field.

As energy is required to distort the orbits, energy is transferred from the electrostatic field to the electrons of each atom between the charged plates. Since energy cannot be destroyed, the energy required to distort the orbits can be recovered when the electron orbits are permitted to return to their normal positions. This effect is analogous to the storage of energy in a stretched spring. A capacitor can thus store electrical energy. An illustration of a simple capacitor and its schematic symbol is shown in Fig. 3-27. The conductors that form the capacitor are called *plates*. The material between the plates is called the *dielectric*. In part B of this figure, the two vertical lines represent the connecting leads. The two horizontal lines represent the capacitor plates. Notice that the schematic symbol (B) and the simple capacitor diagram (A) are similar in appearance. In a practical capacitor, the parallel plates may be constructed in various configurations (circular, rectangular, etc.); but the cross-sectional area of the capacitor plates is tremendously large in comparison to the cross-sectional area of the connecting conductor. This means that there is an abundance of free electrons available in each plate of the capacitor. If the cross-sectional area and plate material of the capacitor plates are the same, the number of free electrons in each plate must be approximately the same. It should be noted that there is a possibility of the difference in charge becoming so large as to cause ionization of the insulating material to occur (cause bound electrons to be freed). This places a limit on the amount of charge that can be stored in the capacitor.

Capacitance is measured in a unit called the farad. This unit is a tribute to the memory of Michael Faraday, a scientist who per-

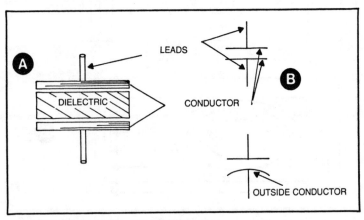

Fig. 3-27. Capacitor and schematic symbols.

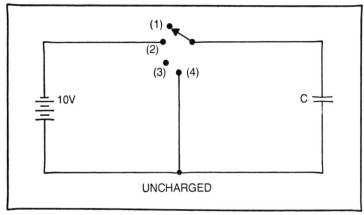

Fig. 3-28. The principle of capacitors charging.

formed many early experiments with electrostatics and magnetism. It was discovered that for a given value of capacitance, the ratio of charge deposited on one plate to the voltage producing the movement of charge is a constant value. This constant value is a measure of the amount of capacitance present. The symbol used to designate a capacitor is (C). The capacitance is equal to one farad when a voltage changing at the rate of one volt per second causes a charging current of one amp to flow. This is expressed by the equation:

$$C = \frac{i}{\frac{\Delta e}{\Delta t}}$$

Where: C = capacitance, in farads
i = instantaneous current in amps
$\frac{\Delta e}{\Delta t}$ = rate of change of voltage, in volts, with time, in seconds

Charging a Capacitor

In order to better understand the action of a capacitor in conjunction with other components, the charge and discharge action of a purely capacitive circuit will be analyzed first. For ease of explanation, the capacitor and voltage source used in Fig. 3-28 will be assumed to be perfect (no internal resistance, etc.), although this is impossible in actual practice.

In Fig. 3-28, an uncharged capacitor is shown connected to a four position switch. With the switch in position 1, the circuit is open and no voltage is applied to the capacitor. Initially, each plate of the capacitor is a neutral body, and until a difference of potential is impressed across the capacitor, no electrostatic field can exist between the plates.

To charge the capacitor, the switch must be thrown to position 2, which places the capacitor across the terminals of the battery. Under the given conditions, the capacitor would reach full charge instantaneously. However, the charging action will be spread out over a period of time in this discussion so that a step-by-step analysis can be made.

In Fig. 3-29, at the instant the switch is thrown to position 2, a displacement of electrons will occur simultaneously in all parts of the circuit. This electron displacement is directed away from the negative terminal and toward the positive terminal of the source. An ammeter connected in series with the source will indicate a brief surge of current as the capacitor charges.

If it were possible to analyze the motion of individual electrons in this surge of charging current, you would see the following action (see Fig. 3-30). At the instant the switch is closed, the positive terminal of the battery extracts an electron from the bottom conductor and the negative terminal of the battery forces an electron into the top conductor. At this same instant, an electron is forced into the top plate of the capacitor and another is pulled from the bottom plate. Thus, in every part of the circuit, a clockwise displacement of electrons occurs in the manner of a chain reaction.

As electrons accumulate on the top plate of the capacitor and others depart from the bottom plate, a difference of potential develops across the capacitor. Each electron forced onto the top plate

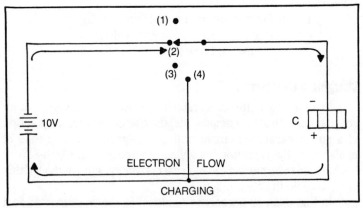

Fig. 3-29. Chart showing the current flow in the circuit during the capacitor's charge sequence.

64

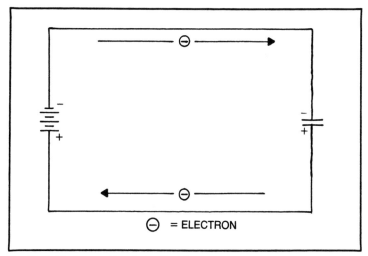
Fig. 3-30. Electron motion during the charge.

makes that plate more negative, while each electron removed from the bottom causes the bottom plate to become more positive. Notice that the polarity of the voltage which builds up across the capacitor is such as to oppose the source voltage. The source forces current around the circuit of Fig. 3-30 in a clockwise direction. The electromotive force (EMF) developed across the capacitor, however, has a tendency to force the current in a counterclockwise direction, opposing the source. As the capacitor continues to charge, the voltage across the capacitor rises until it is equal in amount to the source voltage. Once the capacitor voltage equals the source voltage, the two voltages balance one another and current ceases to flow in the circuit.

In studying the charging process of a capacitor, it must be emphasized that no current flows through the capacitor. The material between the plates of the capacitor must be an insulator.

If you were an observer stationed at the source or along one of the circuit conductors, the action would have all the appearances of a true flow of current even though the insulating material between the plates of the capacitor prevents having a complete path. The current which appears to flow in a capacitive circuit is called *displacement current*.

To provide a better understanding of charging action, a capacitor can be compared to the mechanical system in Fig. 3-31. Part A of the diagram shows a metal cylinder containing a flexible rubber membrane which blocks off the cylinder. The cylinder is then

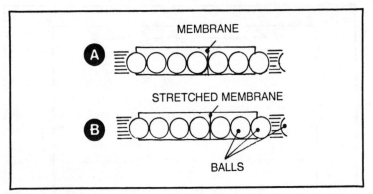

Fig. 3-31. Mechanical equivalent of a capacitor.

filled with round balls as shown. If an additional ball is now pushed into the left hand side of the tube, the membrane will stretch and a ball will be forced out of the right hand end of the tube. Since you can not see inside the tube, the ball will have the appearance of traveling all the way through the tube. For each ball inserted into the left hand side, one ball will leave the right hand side, although no balls actually pass all the way through the tube.

As more balls are forced into the tube, it becomes increasingly difficult to force in additional balls due to the tendency of the membrane to spring back to its original position. If too many balls are forced into the tube, the membrane will rupture, and any number of balls can then be forced all the way through the tube.

A similar effect occurs in a capacitor when the voltage applied to the device is too high. If an excessive amount of voltage is applied to a capacitor, the insulating material between the plates will break down and allow a current flow through the capacitor. In most cases, this destroys the capacitor, necessitating its replacement.

When a capacitor is fully charged and the source voltage is equaled by the *counter electromotive force* (CEMF) across the capacitor, the electrostatic field between the plates of the capacitor will be maximum. Since the electrostatic field is maximum, the energy stored in the dielectric will be maximum.

If the switch is now opened, as shown in Fig. 3-32, the electrons on the upper plate are isolated. Due to the intense repelling effect of these electrons, no electrons will return to the positive plate. Thus, with the switch in position 3, the capacitor will remain charged indefinitely. At this point, note that the insulating dielectric material in a practical capacitor is not perfect and a small leakage current will flow through the dielectric. This current will eventually dissipate the charge. A high quality capacitor may hold its charge for a month or more.

To review briefly, when the capacitor is connected across a source, a surge of charging current will flow. This charging current develops a CEMF across the capacitor which opposes the applied voltage. When the capacitor is fully charged, the CEMF will be equal to the applied voltage and charging current will cease. At full charge, the electrostatic field between the plates is at maximum intensity and the energy stored in the dielectric is maximum. If the charged capacitor is disconnected from the source, the charge will be retained for some period of time. The length of time the charge is retained depends on the amount of leakage current present. Since electrical energy is stored in the capacitor, a charged capacitor can act as a source.

Discharging a Capacitor

To discharge a capacitor, the charges on the two plates must be neutralized. This is accomplished by providing a conducting path between the two plates (see Fig. 3-33). With the switch in position 4, the excess electrons on the negative plate can flow to the positive plate and neutralize its charge. When the capacitor is discharged, the distorted orbits of the electrons in the dielectric return to their normal positions and the stored energy is returned to the circuit. It is important to note that a capacitor does not consume power. The energy the capacitor draws from the source is recovered when the capacitor is discharged.

Factors Affecting Capacitors—Plate Area

To investigate the relationship between capacitance and plate area, the action of two capacitors having different plate areas will be

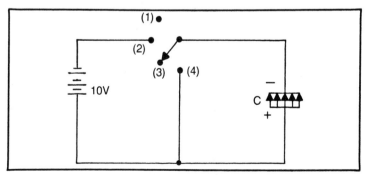

Fig. 3-32. Isolation of the charged capacitor from the circuit.

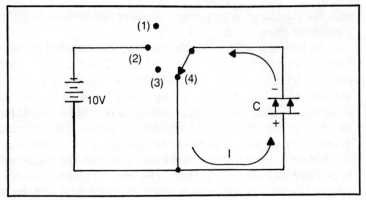

Fig. 3-33. Discharging the capacitor.

compared. Figure 3-34 shows a relatively small capacitor to which a potential of 10 volts is applied. As shown in this diagram, three electrons have been forced onto the top plate of the capacitor by the negative terminal of the battery. Since the area of this plate is small, the electrons are crowded together and repel each other (and

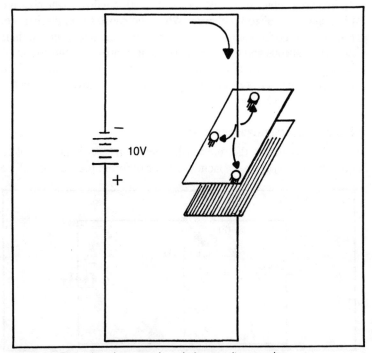

Fig. 3-34. Plate size plays a major role in capacitance value.

additional electrons from the source as well) with great force. Due to this repelling effect, the source has difficulty in forcing additional electrons into the capacitor. Since capacitance is the ratio of charge stored to voltage applied (C = Q/E), this capacitor does not have much capacitance. If the plate area is large, such as in Fig. 3-35, the electrons can spread out over the plate, and the pressure (CEMF) developed by a given number of electrons is small. A given value of source voltage can therefore force more electrons onto the large plate than onto the small plate. Thus, the capacitor with the larger plate area has a greater capacity for storing charge. Capacitance is directly proportional to plate area. Doubling the plate area will double the capacitance.

Plate Spacing

If the distance between the plates of a capacitor is changed, the capacitance will change. As the distance between the plates of a capacitor is reduced, the capacitance will increase. This can be

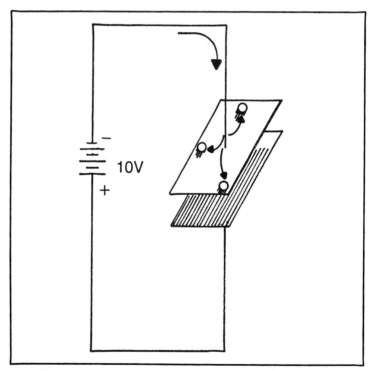

Fig. 3-35. A pictorial representation of the effect created by a large capacitor plate area.

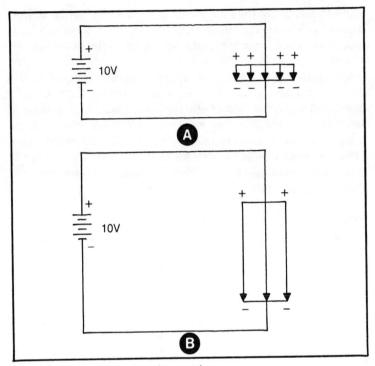

Fig. 3-36. Effects of capacitor plate spacing.

explained by an examination of Fig. 3-36, which shows two capacitors having equal plate areas and unequal plate spacing.

In Fig. 3-36A, the distance between the plates is small. As a result of the close spacing of the plates, the positive charges on the top plate partially neutralize the field surrounding the negative charges on the bottom plate and the source can force more electrons onto the bottom plate. Similarly, the field about the positive charges on the top plate is partially neutralized by the electrons on the bottom plate and the source can remove more electrons from the top plate. When the plates are far apart, very little neutralization of field takes place and the source cannot place as much charge on the capacitor. This example illustrates the fact that capacitance is inversely proportional to plate spacing. The greater the distance between the plates, the smaller will be the capacitance.

Dielectric Material

The dielectric material is the insulating material used between the capacitor plates. The amount of capacitance contained by a pair of plates is affected to a great degree by the type of dielectric material inserted between the plates. Through experimentation, scientists discovered that a given set of plates exhibit minimum capacitance when the area between the plates contains a vacuum. If a nonconductor such as glass is inserted between the plates in place of the vacuum, the capacitance will increase. Some modern ceramic materials can produce a capacitance several hundred times greater than that obtained with a vacuum between the same set of plates.

In order to be able to compare dielectric materials as to their ability to increase capacitance, a number is assigned to each dielectric material. This number, called the *dielectric constant*, tells how many times the material can increase the capacitance as compared to a vacuum dielectric.

The dielectric constants of several materials are listed in Fig. 3-37. Notice the dielectric constant for a vacuum. Since a vacuum is the standard of reference, it is assigned a constant of one; and the dielectric constants of all materials are compared to that of a vacuum. Since the dielectric constant of air has been determined experimentally to be approximately the same as that of a vacuum, the dielectric constant of air is also considered to be equal to one. The formula used to compute the value of capacitance using the physical factors just described is:

$$C = 0.2249 \left(\frac{KA}{d} \right)$$

where: C = capacitance, in picofarads (10^{-12})
A = area of one plate in square inches
d = distance between plates in inches

MATERIAL	CONSTANT (K)
VACUUM	1.0000
AIR	1.0006
PARAFFIN PAPER	3.5
GLASS	5 - 10
MICA	3 - 6
RUBBER	2.5 - 35
WOOD	2.5 - 8
GLYCERINE (15°C)	56
PETROLEUM	2
PURE WATER	81

Fig. 3-37. Chart showing the dielectric constant of various materials.

K = dielectric constant of insulating material
0.2249 = a constant resulting from conversion from Metric to British units.

TYPES OF CAPACITORS

Capacitors are classified into two general types: *variable* and *fixed*. Variable capacitors are those that are constructed in a fashion that allows the value of capacitance to be varied over a prescribed range. There are two types of variable capacitors: the *rotor-stator* type and the *trimmer* type.

The rotor-stator types use air as the dielectric. The amount of capacitance is varied by changing the position of the rotor plates (movable plates). This changes the effective plate area of the capacitor. When the rotor plates are fully meshed between the stator plates, the capacitance is maximum. The rotor-stator type is illustrated in Fig. 3-38.

The trimmer type of variable capacitor consists of two plates separated by a dielectric other than air. The capacitance is varied by changing the distance between the plates. This is ordinarily accomplished by means of a screw which forces the plates closer together. This type of variable capacitor is shown in Fig. 3-39.

Fixed capacitors are categorized by the type of dielectric used. The following is a description of some of the more common types of fixed capacitors.

Paper Capacitor

A paper capacitor is one that uses paper as its dielectric. The construction of the typical paper capacitor is shown in Fig. 3-40. It

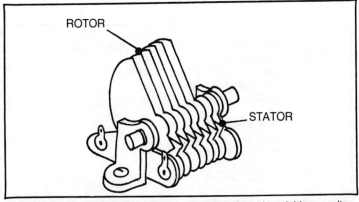

Fig. 3-38. Pictorial and schematic representation of an air variable capacitor.

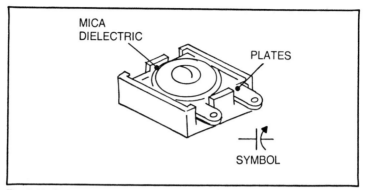

Fig. 3-39. Trimmer capacitor.

consists of flat, thin strips of metal foil conductors separated by the dielectric material. In this capacitor, the dielectric used is waxed paper. Paper capacitors usually range in value from about 300 picofarads to about 4 microfarads. Normally, the voltage limit across the plates rarely exceeds six hundred volts. Paper capacitors are sealed with wax to prevent the harmful effects of moisture.

Mica Capacitor

Mica capacitors consist of alternate layers of mica and plate material. Their capacitance is of a small value, usually in the picofarad range. Although small in physical size, the mica capacitors have a high voltage handling capacity. Figure 3-41 shows how mica capacitor values are indicated by use of colored dots on the case of the capacitor.

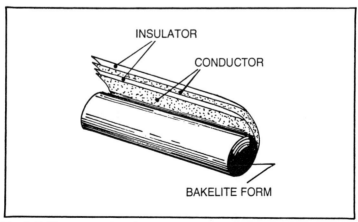

Fig. 3-40. The construction of a typical paper capacitor.

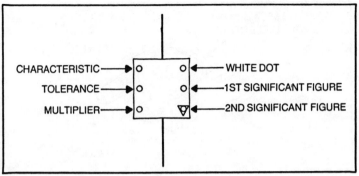

Fig. 3-41. Conventional molded mica capacitors are color-coded with colored dots.

Oil Capacitor

Oil capacitors are often used in radio transmitters where high output power is desired. Oil-filled capacitors are nothing more than paper capacitors that are immersed in oil. The oil-impregnated paper has a high dielectric constant which lends itself well to the production of capacitors that have a high value. Many capacitors will use oil with another dielectric material to prevent arcing between the plates. If an arc should occur between the plates of an oil-filled capacitor, the oil will tend to reseal the hole caused by the arc. These types are often called *self-healing capacitors*.

Ceramic Capacitor

Ceramic capacitors are so-named because of the use of ceramic dielectrics. One type of ceramic capacitor uses a hollow ceramic cylinder as both the form on which to construct the capacitor and as the dielectric material. The plates consist of thin films of metal deposited on the ceramic cylinder.

A second type of ceramic capacitor is manufactured in the shape of a disk. After leads are attached to each side of the capacitor, the capacitor is completely covered with an insulating moisture-proof coating. Ceramic capacitors usually range in value between one picofarad and 0.01 microfarad and may be used with voltages as high as thirty thousand volts.

Electrolytic Capacitors

Electrolytic capacitors are used where a large amount of capacitance is required. As the name implies, electrolytic capacitors contain an electrolyte. This electrolyte can be in the form of either a liquid (wet electrolyte) or a paste (dry electrolyte). Wet electrolytic capacitors are no longer in popular use due to the care needed to prevent spilling of the electrolyte.

Dry electrolytic capacitors consist essentially of two metal plates between which is placed the electrolyte. In most cases, the capacitor is housed in a cylindrical aluminum container which acts as the negative terminal of the capacitor, as shown in Fig. 3-42. The positive terminal (or terminals if the capacitor is of the multisection type) is in the form of a lug on the bottom end of the container. The size and voltage rating of the capacitor is generally printed on the side of the aluminum case.

Internally, the electrolytic capacitor is constructed similarly to the paper capacitor. The positive plate consists of aluminum foil covered with an extremely thin film of oxide which is formed by an electrochemical process. This thin oxide film acts as the dielectric of the capacitor. Next to and in contact with the oxide is a strip of paper or gauze which has been impregnated with a paste-like electrolyte. The electrolyte acts as the negative plate of the capacitor. A second strip of aluminum foil is then placed against the electrolyte to provide electrical contact to the negative electrode (electrolyte). When the three layers are in place, they are rolled up into a cylinder, as shown at the bottom of Fig. 3-43.

Electrolytic capacitors have two primary disadvantages, in that they are polarized and they have a low leakage resistance. Should the positive plate be accidentally connected to the negative terminal of the source, the thin oxide film dielectric will dissolve and the capacitor will become a conductor (i.e., it will short). The polarity of the terminals is normally marked on the case of the capacitor. Since electrolytic capacitors are polarity sensitive, their use is ordinarily restricted to dc circuits or circuits where a small ac voltage is superimposed on a dc voltage. Special electrolytic capacitors are available for certain ac applications, such as motor starting capacitors. Dry electrolytic capacitors vary in size from about four microfarads to several thousand microfarads and have a voltage limit of approximately five hundred volts.

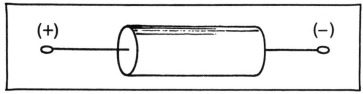

Fig. 3-42. Electrolytic capacitor.

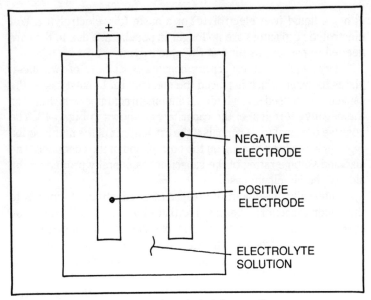

Fig. 3-43. Internal construction of an electrolytic capacitor.

The type of dielectric used and its thickness govern the amount of voltage that can safely be applied to a capacitor. If the voltage applied is high enough to cause the atoms of the dielectric material to become ionized, and arc over will take place between the plates. If the capacitor is not self-healing, its effectiveness will be impaired. The maximum safe voltage of a capacitor is called its *working voltage* and is indicated on the body of the capacitor. The working voltage of a capacitor is determined by the type and thickness of the dielectric. If the thickness of the dielectric is increased, the distance between the plates is also increased and the working voltage will be increased. Any change in the distance between the plates will cause a change in the capacitance of a capacitor. Because of the possibility of voltage surges (brief high amplitude pulses), a margin of safety should be allowed between the circuit voltage and the working voltage of a capacitor. The working voltage should always be higher than the maximum circuit voltage.

COLOR CODES FOR CAPACITORS

Although the value of a capacitor may be indicated by printing on the body of the device, many capacitive values are indicated by the use of a color code. The colors used to represent the numerical value of a capacitor are the same as those used to identify resistance values. There are two color coding systems that are currently in popular use: The Joint Army-Navy (JAN) and the Ratio Manufacturers' Association (RMA) system.

In each of these systems, a series of colored dots (sometimes bands) is used to denote the value of the capacitor. Mica capacitors are marked with either three dots or six dots. Both systems are similar, but the six dot system contains more information about the capacitor, such as working voltage, temperature coefficient, etc. Capacitors are manufactured in various sizes and shapes. Some are small tubular resistor-like devices, and others are molded, rectangular, flat components. Figure 3-44 shows some of the more common shapes of capacitors.

CAPACITORS FOR POWER SUPPLIES

The previous discussion about capacitors was presented to give the reader a general idea of how capacitors function and the many types available. While small value capacitors may be used in some power supply circuits for protection of solid-state rectifiers and some noise elimination problems, we are mostly interested in the larger types which will be used for filter applications. In modern power supplies, these are almost always electrolytic capacitors, although some of the larger oil-filled types may also be substituted. An electrolytic capacitor is able to provide a very large capacitance value in a physically small device. Oil-filled capacitors are physically larger than are most electrolytics and are limited in capacitance value.

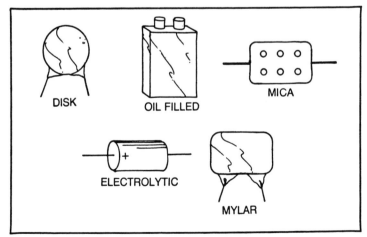

Fig. 3-44. Common capacitor shapes.

Oil-filled capacitors do have one distinct advantage, especially when they are used in medium and high voltage power supplies. These devices usually have higher dc voltage ratings than do electrolytics. The dc voltage rating is most important, in that the capacitor must be able to withstand the full dc potential of the power supply output. Practical design considerations usually call for the dc voltage rating to be about 25 percent higher than the maximum anticipated dc output from the rectifiers. For instance, a 12-volt power supply should use a filter capacitor which is rated for at least 15 volts dc. Again, we are talking here about the maximum output from the rectifiers that will be seen across the capacitor terminals. The output from the rectifiers is pulsating dc, whose average value may be measured with a dc voltmeter. The capacitor, however, will charge to a potential equal to the *peak* dc value, which will be about 1½ times the measured voltage at the rectifier. Figuring capacitor working voltage would involve determining the expected output from the rectifiers, multiplying this figure by 1½ times, and then tacking on an additional 25 percent or so for safety measures.

The value of the capacitor in microfarads is usually non-critical above a certain point. Generally speaking, the higher the capacitance rating, the better the power supply will perform. Here, we are concerned about dynamic regulation. A simple power supply will tend to present voltage swings in relationship to the amount of current which is drawn by the load. The voltage will be high when current drain is low and will drop significantly when drain is high. A power supply with good dynamic regulation will have far less voltage swing than will one with poor dynamic regulation.

In simple power supplies, the filter capacitor will determine the amount of dynamic regulation the entire circuit exhibits. Low voltage power supplies may require filter capacitance values of a thousand microfarads or more. These high values are quite easy to attain using relatively small components. But in moderate to high voltage dc supplies, these values are practically impossible to get, at least while staying within the realm of practicality. As the working voltage rating of the capacitor increases, the physical size increases proportionally. A 1,000 microfarad capacitor with a working voltage of 12 Vdc can easily be held in the palm of the hand. The same capacitance value at 3,000 Vdc would weigh several hundred pounds. Fortunately, high voltage power supplies do not generally require dynamic regulation in excess of what can be obtained with 20 to 30 microfarad filter capacitors.

The power supply designer and builder must pay strict attention to the voltage ratings of all components, especially filter capacitors. When electrolytic capacitors are operated high above their working voltage limits, the internal electrolyte will begin to heat and expand. While blow-out safety plugs are often included with these devices, they don't always work. Electrolytic capacitors can explode and cause serious injury to bystanders. Other types of capacitors generally tend to short-circuit when subjected to excess voltage levels, but an explosive potential can still exist with oil-filled varieties. One other note of caution: Filter capacitors are generally dc devices. When electrolytics are connected to an ac source of considerable potential, they can overheat and explode.

It was mentioned in the earlier discussion that unlike most capacitors, electrolytics are almost always polarized devices. This means that there is a negative terminal and a positive one. There are some non-polarized electrolytic capacitors on the market, but these are not used in power supply applications. Damage to electrolytic capacitors will result if they are incorrectly connected in regard to polarity. Also, the power supply will not work properly.

CAPACITORS IN PARALLEL

Like many other types of electronic devices, several capacitors may be combined to increase overall ratings. If a particular circuit requires a capacitance value of 1,000 microfarads at 50 volts dc, two 500 microfarad units may be combined in parallel to arrive at the desired value. This circuit is shown in Fig. 3-45. Here, each capacitor must have a working voltage rating of 50 Vdc and the same capacitance value. Capacitors in parallel add capacitance values. Two 500 microfarad components, each rated at 50 Vdc, and wired in parallel will deliver an overall value of 1,000 microfarads at 50 Vdc. Three of these units wired in parallel would give you 1,500 mi-

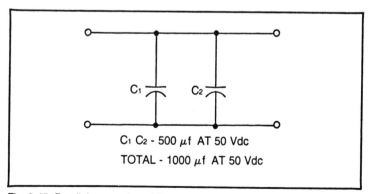

Fig. 3-45. Parallel connection of filter capacitors.

crofarads, still at 50 Vdc. Using the parallel connection method, several small capacitors can serve to form a moderately large one.

When interconnecting capacitors, it is desirable for all units in the combination circuit to be of the same value and preferably made by the same manufacturer. This recommendation is made not so much from an operational standpoint but due to the fact that matching capacitors are usually more easily wired and mounted together in uniform fashion. Actually, you may easily combine a capacitor rated at 50 microfarads, another with a 10 microfarad rating, and still a third rated at 100 microfarads. It is important that each capacitor in this parallel circuit have a working voltage rating high enough to withstand the dc potential. Using the components just mentioned, the total capacitance value would be 160 microfarads. The voltage rating would be equal to the lowest rating of any of the three capacitors. For example, if the 50 microfarad unit had a 20 Vdc rating, the 100 microfarad device, 6 Vdc, and the 10 microfarad capacitor a 450 volt rating, then the working voltage rating of the circuit would be 6 Vdc, the lowest value of the three components. When electrolytic capacitors are used in parallel circuits, be certain that all polarities match. A correct and an incorrect circuit is shown in Fig. 3-46.

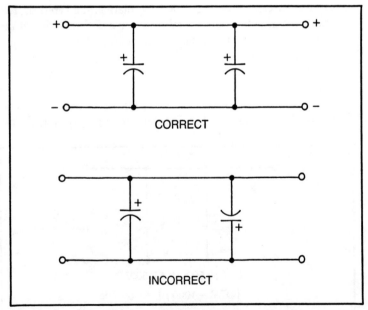

Fig. 3-46. Parallel connection of electrolytic capacitors requires matching polarities.

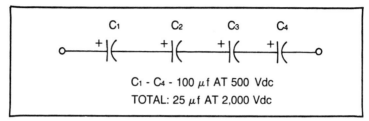

Fig. 3-47. Four capacitors in a series string.

CAPACITORS IN SERIES

In moderate and high voltage power supplies, it is often difficult to obtain filter capacitors with working voltage ratings adequate to handle the dc potential. Electrolytic capacitors normally have maximum ratings of 450 Vdc, although a few small capacitance units may be rated as high as 600 Vdc. Oil-filled capacitors are often rated at 1,500 Vdc or more, and some specialized units may be ordered which will withstand a dc potential of over 5,000 Vdc. Most of these high voltage capacitors have rather low capacitance values and must be connected in parallel to obtain sufficient dynamic regulation. The main drawbacks of high voltage oil-filled capacitors are their size, weight, and price.

It is possible to combine several capacitors in series to increase the overall working voltage value. Figure 3-47 shows a series string of four capacitors with individual ratings of 100 microfarads at 500 Vdc. When wired in series as shown, the working voltages add. So, four 500 Vdc units in series gives us a total working voltage of 2,000 Vdc. But now the drawback comes to the front. While the working voltage adds with each unit, the capacitance value *divides*! To arrive at the total circuit capacitance, you must divide the microfarad rating of one capacitor by the total number of capacitors in the series string. In our sample circuit, this would work out to 100 microfarads divided by 4 units, or 25 microfarads.

To have an effective series string, it is necessary to select individual capacitors which have high microfarad ratings and dc voltage ratings which are also as high as possible. The higher voltage ratings mean that less capacitors will be required to obtain the desired voltage value, and the higher capacitance figures will leave you with a higher overall capacitance value after the division process has taken place.

Electrolytic capacitors are often connected in series strings for application in medium and high voltage power supply circuits. Electrolytic capacitors are ideal, because they are able to produce a high capacitance in a physically small package. Figure 3-48 shows how eight 160 microfarad electrolytic capacitors rated at 450 Vdc each are combined in series to produce a complex capacitor rated at 3,600 Vdc and with a capacitance value of 20 microfarads. This is a conventional series capacitor string which is used in many power supply circuits rated to deliver about 3,000 Vdc at the output. The dynamic regulation here is quite good. Notice that each electrolytic capacitor is connected in strict observance of polarity. The positive terminal of the first unit is connected to the negative terminal of the next. This latter capacitor has a connection made from its positive terminal to the negative terminal of the next, and so on. This differs from the parallel connection where all positive terminals are connected together at one circuit point and all negative terminals at another.

This type of series circuit mandates the use of identical capacitors. They must all be of the same capacitance and voltage ratings and a product of the same manufacturer. In this circuit, each capacitor will drop one-eighth of the total output voltage . . . but only if they are perfectly matched. If this circuit were used with a power supply which produced a maximum output of 3,000 Vdc, then each capacitor would be expected to withstand a potential of 3,000 Vdc divided by eight (the number of capacitors in the string). This gives us a value of 375 Vdc. Since each capacitor is rated to withstand 450 Vdc, an adequate safety margin is built in. But, if non-matching units were used, those capacitors with higher capacitance values would drop more voltage. This might mean that other capacitors in the string would only have 150 or 200 Vdc present at their terminals, while the larger capacitance units would see a potential of 550 volts or more. This is in excess of their 450 volt rating and destruction could occur. Even though all capacitors used may have the same printed capacitance and voltage ratings, these can be off by 10 to 20 percent. This especially applies when apparently identical units from two or more manufacturers are used. One manufacturer's capacitor may actually have a value of 150 microfarads, while another's may be producing 170 microfarads. Again, all of these units are rated at a nominal 160 microfarads. Due to the inequalities of capacitance values, one capacitor may assume a voltage value in excess of its working voltage rating and be destroyed.

This situation is usually prevented by using all capacitors of the same type and from the same manufacturer. By incorporating all capacitors of the same manufacturing series, their capacitance rat-

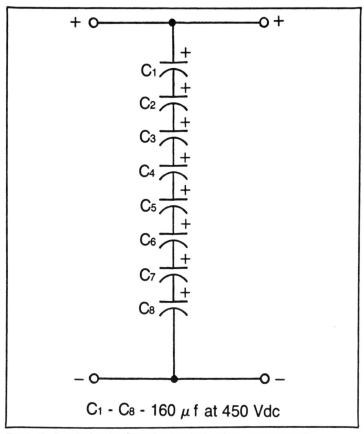

Fig. 3-48. A typical series filter circuit in high-voltage power supplies.

ings should be far better matched. It is possible in some instances to individually match similar capacitors from several manufacturers by installing matching resistors at their terminals, but this is a trial and error process that can take many hours and still not provide proper results.

Even when matching units are combined in a series circuit, it is necessary to further assure this match by installing moderately large value resistors across each terminal. Fortunately, all of these resistors can be of the same value and individual matching is not necessary as was discussed above. The latter process would mean that resistors of different ohmic values would be needed at each terminal.

Figure 3-49 shows the former series string with 25,000 ohm resistors installed across each capacitor. These perform the same

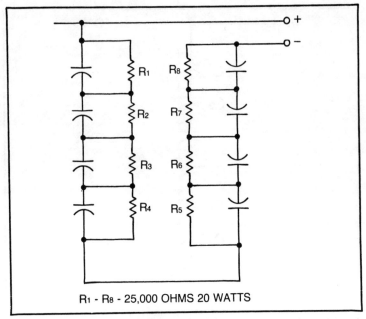

Fig. 3-49. Parallel resistors are usually used in a series capacitor strings for equalization and to bleed away the stored charge.

function as the parallel resistors in an earlier discussion on matching series rectifiers. This protective circuit is quite standard among builders of high voltage power supplies and serves two purposes. First of all, the resistors provide a better match from capacitor to capacitor. Secondly, they form a complex bleeder resistor which is rated at 200,000 ohms (25,000 ohms times 8 resistors). In this particular circuit, each resistor is rated at 20 watts. The bleeder resistor serves to stabilize the output voltage of the supply and discharges the capacitors when the circuit is deactivated.

This is a good point to talk about safety, especially as it applies to moderate and high voltage power supplies. There is no such thing as a slight electrical shock when voltage potentials rise above about 200 volts. This goes triple when talking about potentials in excess of 1,000 Vdc. Severe burning and mutilation can occur in an instant when someone comes in contact with high voltage. I vividly remember viewing the foot of a fellow experimenter who had come into contact with a 3,000 volt supply which was floor mounted. Fortunately, he survived, but his right foot and a portion of his ankle looked like it had been chewed on by some large carnivore. What may be startling to some readers is the fact that this power supply had been turned off and the power cord removed from the wall outlet a full five minutes before his accident occurred. How can this happen? *Capacitors store electrical energy*. If they are not discharged, this potential is still in storage and will be for quite some time. The bleeder resistor circuit described above is designed to quickly discharge the capacitors after power is removed from the transformer. But what if the bleeders fail? This is exactly what happened to my friend. The power supply upon which he was working had been altered during the servicing procedure so that the bleeder resistors were not connected to the capacitor terminals. The same electrical potential was present at the unconnected output of the supply five minutes after the cord was pulled from the wall that was there when the supply was originally turned on.

You may have noticed that some of the larger capacitors that come in factory-sealed cartons from the manufacturer have a shorting wire connected across their terminals. This assures that there is no stored energy in the component. This is especially true of high voltage capacitors, which are charged to full potential during the manufacturer's checkout procedure. Some stories have been told of persons being killed by these new devices while they were being removed from their packing cartons. This is highly unlikely and I have never been able to substantiate this oft-talked about story, but it is possible. Always be suspicious of any capacitor. It is a simple procedure to grasp the capacitor by its insulated case and to short out the electrode leads with a large *insulated* screwdriver. It's always a good idea to store unused high voltage capacitors with these terminals shorted together by a piece of wire. I was mildly shocked several years ago by a capacitor which was discharged and stored in an attic. This occurred shortly after an electrical storm, so it would appear that the static charge in the atmosphere also charged the high value capacitor to a low level. This shock was very minor, but my reflex reaction caused me to jump backwards, striking my head and ending up with a mild concussion. This goes to prove that serious accidents can occur from even a mild electrical "tickle" if it is unexpected.

Many technicians and engineers will disagree with this next statement, but I firmly believe in it and practice it daily. "It is not enough to have a healthy respect for high voltage; it is mandatory that high voltage be feared." True, this is a controlled fear which is necessary if high voltage supplies are to be built and serviced, but nevertheless, it is a fear. The person who does not fear the consequences of coming in contact with high voltage is a fool.

Again, many argue that knowledge of high voltage circuits should assuage the fear element, but this is not enough. Certainly, it is mandatory that you know where all high voltage contact areas lie, but you must remember that electronic circuits are often highly complex and unpleasant surprises can often be had. This is especially true when working on defective equipment. Damaged components can cause high voltage to appear in highly unlikely places. Nevertheless, it is there and can easily kill you. The initial checkout of a newly built high voltage power supply is another area where similar problems may occur due to unknown wiring errors.

Another potentially hazardous situation can develop when measuring high voltage potentials, especially when the instrument used is an inexpensive multimeter. Many of these devices will measure dc potentials of 3,000 volts or more, but often, the insulation on a test lead is rated for 1,000 volts or less. This means that if your body is well grounded, you could get a life-threatening shock through the insulation. Also, the rubber insulation on these test leads is often nicked by pliers and screwdrivers and is quite commonly burned through in some places by coming in too close contact with a heated soldering iron. These may seem like minor problems, but it takes only one to kill you.

Too many persons tend to relax their fears a bit when working on low voltage power supplies. While it is true that there is little danger from the output of a 6 or 12 volt dc power supply, the input which is usually derived from the ac line poses many hazards. More people are killed and injured by household 115 Vac contact than any other voltage source. This is exactly what is present at the primary input to your power supply transformer. Even with the switch in the off position, this 115 volt potential is still present at one switch contact, at the fuse holder, and possibly at several other places in the primary circuit. This is simple to overcome and only involves unplugging the line cord.

Safety is discussed at this midway point in a chapter on electronic components because filter capacitors in some power supplies offer deadly potentials to the builder. But this is not the only area where safety hazards exist. When working with any power supply that is driven by the ac line, a thorough understanding of the circuit and of proper safety procedures is mandatory.

CAPACITOR INSULATION

Having discussed power supply safety, especially in regard to stored power in filter capacitors, we will return now to the filter

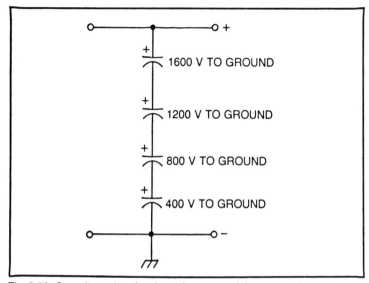

Fig. 3-50. Capacitor string showing voltage potentials to ground.

capacitor string. Figure 3-50 shows another simple series string composed of four electrolytic capacitors. Let's assume that the voltage rating of each component is 400 Vdc. This means that the total working voltage of the string is 1,600 Vdc. Also assume that the maximum value of output voltage from this string when connected to the remainder of the power supply circuitry is 1,200 Vdc. Capacitor insulation now becomes a prime factor. Remember, each capacitor is insulated to withstand a working voltage of 400 Vdc. With a 1,200 Vdc power supply output, each unit will drop about 300 volts. But what is the potential between the first capacitor and circuit ground? The answer is 1,200 Vdc. Remember, there are three other capacitors between this first one and actual circuit ground. But if the outer case of the first capacitor should come in contact with chassis ground, then this component's insulation rating would be exceeded. An arc could occur between the positive electrode and the capacitor case, destroying the unit. Again, the capacitor is only dropping 300 volts, but if its metal case is connected to circuit ground, a potential difference of 1,200 Vdc exists between the positive electrode and the metal case, which is at ground potential. This assumes that the metal case does not connect to the negative electrode within the capacitor, although this is often the case. If this is the situation, then an immediate short circuit would exist.

Many electrolytic capacitors which use the metal case as the negative electrode contact contain external insulation which may be in the form of paper or flexible plastic. There is a good chance that these materials will not effectively insulate the capacitor body from chassis ground. This is especially true at high voltage potentials.

This situation is easily overcome by installing all capacitors in the series string on a piece of plexiglass or some other form of insulating material. The last capacitor in the string only sees a 300 volt potential to chassis ground, so this unit need not be given special attention and may be mounted directly to the chassis. The second to the last capacitor, however, would see a 600 volt potential to chassis ground and should be isolated above ground with the rest.

From a safety standpoint, it is quite dangerous to allow your hand to come in contact with the insulated bodies of any of the capacitors in this string while they are charged. If you were to touch the insulated case of the top capacitor with one hand while resting your other hand on the power supply chassis, a potential current path exists between your two hands and through your chest cavity. An electrical shock through the heart can temporarily stop its action altogether. If no one is near to administer external massage, this temporary stoppage often becomes permanent.

While most of the power supply projects in this book do not include deadly output potentials, some do. The discussion on capacitors and power supply safety is in no way meant to frighten you to a point where you no longer want anything to do with power supply circuits. Even high voltage power supplies can be safely built, serviced, and maintained; but knowledge of what can happen when caution is not exercised is mandatory.

BLEEDER RESISTORS

Bleeder resistors have already been discussed in some detail. However, a bit more knowledge is necessary about these devices which are so appropriately named. The main purpose of a bleeder resistor is to bleed off stored current in the filter capacitor or capacitor bank. A bleeder resistor is normally a fixed resistor which is connected across the output terminals of the power supply. For most applications, the bleeder resistor should have an ohmic value equivalent to the dc output voltage times 100. This would mean a 100 volt dc output supply might use a 10,000 ohm bleeder resistor. The power rating of the chosen bleeder will vary, depending upon the total ohmic value and the dc output voltage. Using the previous example of a 10,000 ohm resistor across the output terminals of a

100 volt supply, we must use Ohm's Law to figure the amount of power which will be dissipated in this component. The derivative of Ohm's Law used here is expressed as $P = \dfrac{E^2}{R}$. This means that power dissipation (P) is equal to the output voltage squared (E^2) divided by the bleeder resistance (R). Inserting 100 volts for E and 10,000 ohms for R, we arrive at:

$$P = \frac{100^2}{10,000} \text{ or } P = \frac{10,000}{10,000} \text{ or } P = 1 \text{ watt}$$

We now know that the 10,000 ohm bleeder resistor must be able to withstand a dissipation of 1 watt. But remember, a bleeder resistor is a safety device. Without it, accidental shock can occur. We know from the above formula that this resistor will be dissipating the maximum amount of power it is rated to withstand during every minute of power supply operation. We have no safety factor built in here. In this case, it would be wise to use a 2 watt resistor, or even better, a 5 watt unit. This allows for a substantial safety factor which is even more important in moderate and high voltage power supplies.

As another example, let's use a power supply with an output of 3,000 Vdc. At 100 ohms per volt, a bleeder resistor with an ohmic value of approximately 300,000 would be required. Now, let's figure the amount of power this device will be required to dissipate. Using the formula from before, we arrive at:

$$P = \frac{3,000^2}{300,000} \text{ or } P = \frac{9,000,000}{300,000} \text{ or } P = 30 \text{ watts}$$

During regular operation of this high voltage power supply, the bleeder resistor would have to dissipate 30 watts continuously. Here, we would choose a bleeder resistor with at least three times this rating and maybe even more. A 100 watt unit would be appropriate, although 150 to 200 watt bleeder resistors are quite common with this type of supply.

Unfortunately, high wattage resistors are almost always of wire wound construction. Should the device heat up, a wire element may break. This effectively removes the resistor from the circuit entirely. Danger! When a bleeder resistor opens up, the filter capacitors may hold their charge.

In moderate and high voltage power supplies, it is usually necessary to provide a bit more protection in case of a bleeder resistor failure. Figure 3-51 shows how this is often accomplished.

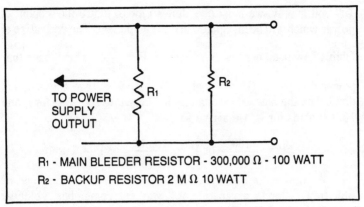

Fig. 3-51. Paralleling the main bleeder resistor with a high value backup unit assures capacitor discharge.

The main bleeder resistor with the 300,000 ohm, 100 watt rating is wired across the power supply output. In parallel with this device is a carbon composition resistor which serves as a backup. This will usually have a very high ohmic value (on the order of 2,000,000 ohms and above) and would have a rating of about five to ten watts. Its large size has little effect on the overall circuit when the main bleeder is functioning properly, but should it open, the auxiliary handles the full job of discharging the capacitors. Due to its high ohmic value, the discharge rate will be slower than with the main bleeder; but complete discharge should be had within thirty seconds.

Bleeder resistors tend to operate at fairly high temperatures, so they should be examined periodically for any signs of damage. If a defective resistor is found, the power supply should not be operated again until a replacement part has been installed. Warning: Do not depend upon the auxiliary bleeder for full time service should the main unit open up. The auxiliary should be thought of as emergency standby only.

Some high voltage power supply circuits take special precautions to assure proper operation of bleeder resistors. Here, a small fan is often used to direct a steady stream of cooling air across these components. This results in cooler operation and a longer dependable life.

SEMICONDUCTOR DEVICES

The solid-state rectifier which has already been discussed is a semiconductor device which is found in nearly every dc power supply. The previous portion of this chapter has dealt with the components found in most unregulated power supply circuits, but these will also be found in even the most complex designs. Electronic regulation usually begins at the output of the filter capacitor of a basic supply and incorporates additional solid-state circuits to improve voltage regulation.

Semiconductor components encompass those devices which are made up of materials that exhibit a conductivity halfway between that of a conductor and an insulator. These devices are made from crystalline materials which are chemically treated with impurities and then combined in sandwich style. The finished crystalline chip is then mounted in a small metal or plastic case through which leads protrude for attachment to the internal structure. Semiconductors contain no moving parts and are not subject to damage from vibration or moderate shocks. Semiconductor devices include many different types of components that perform a myriad of electronic functions. Each will be discussed individually.

Transistors

Transistors are devices that can amplify a signal. In power supply circuits, they are most often used to form series regulators at the output of the filter capacitor. A transistor utilizes a small change in current to produce a large change in voltage, current, or power. The transistor, then, may function as an amplifier or an electronic switch. There are many different types of transistors with individual characteristics, but the theory of operation is basic to all of them.

The advent of the transistor has opened a completely new field for the development of portable equipment. The compactness and ruggedness of transistorized equipment has allowed the manufacture of portable equipment that was previously impractical. Transistors are now being used in mobile equipment, test equipment, recording equipment, photographic equipment, hearing aids, radios, etc. In other words, transistors may be used in almost any application where low and medium power electrons are used.

Figure 3-52 shows pictorial drawings of one type of transistor, the bipolar transistor. There are three leads protruding from this device which are connected to the emitter, collector, and the base of the component. Transistors have the ability to indirectly increase electrical power and this is often rated in terms of multiplication. Shown in Fig. 3-53 are the two basic types of bipolar transistor types, which are labeled PNP and NPN. PNP transistors are composed of a negative semiconductor material (N) sandwiched be-

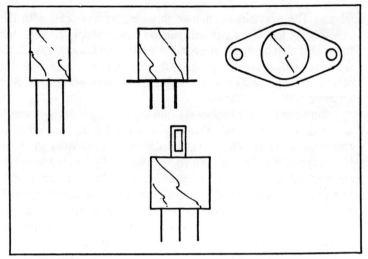

Fig. 3-52. Bipolar transistors come in many different packages.

ween two sections of positive material (P). The reverse is true of the NPN units, which sandwich the P material between two sections of an N material.

Semiconductor devices are described in two ways: limiting conditions and characteristics. Additionally, the manufacturer will usually give each device a manufacturer's type number and a packaging type. This refers to the case the component is mounted in.

Certain symbols will be used as abbreviations for describing characteristics and operating parameters of these devices. While there are many such abbreviations, only the most prevalent ones will be discussed here. The first abbreviation stands for device dissipation. It is abbreviated P_t. This rating is most often expressed in watts, although in certain devices, it may be expressed in milliwatts. P_t describes the maximum amount of power in watts that can be dissipated by the component and still remain within the manufacturer's ratings.

Another important rating is that of collector current, abbreviated I_c. I_c describes the largest amount of current which can be drawn through the collector of the transistor without exceeding manufacturer's ratings. This condition is normally described in amperes of dc current, although it may be described in milliamperes for lower power devices.

The third condition used to describe bipolar transistors is breakdown voltage, and it is expressed in three different forms.

First is the collector to base breakdown voltage, which is abbreviated V_{cvo}. The second abbreviation is V_{ceo}, which is a description of the collector to emitter breakdown voltage. The third and final abbreviation describes the emitter to base breakdown voltage and is displayed as V_{ebo}. Breakdown voltage is the maximum applied voltage to the various portions of the transistor that can be tolerated by the device. When these ratings are exceeded, the device is being operated outside of the manufacturer's ratings and may soon fail to operate.

In addition to limiting conditions, bipolar transistors may also be described as to operating characteristics. One of these characteristics, which is abbreviated h_{FE}, stands for the typical current gain of the device. This is also known as the amplification factor. This figure is not stated in any specific term but uses a single number to indicate the multiplication factor. A second characteristic is the typical gain bandwidth, abbreviated f_T. Generally, this indicates the maximum frequency the transistor will operate while still exhibiting typical characteristics. The typical gain bandwidth may be given in kilohertz (kHz) or in megahertz (MHz).

Finally, most manufacturers will list the type of case or cases in which the device is housed. Certain manufacturers may offer the same device in several different cases. Case style and electronic lead configuration or basing will usually be described with numbers such as TO-1, to TO-92, etc.

Silicon-Controlled Rectifiers

The silicon-controlled rectifier (SCR) is a solid-state device which is made up of layers of crystalline material which has been treated to make it into a semiconductor. They are sometimes used for special control purposes in dc power supplies as switching regulators and in battery chargers. The SCR is a special kind of diode which allows current to pass through the device upon a command from an external source. This signal is applied between

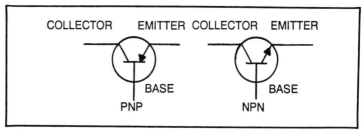

Fig. 3-53. Schematic representation of the two basic types of bipolar transistors.

the anode and the third element, called the gate. Aside from the capability to switch in this manner, the SCR is similar to a standard diode. The SCR's output will be direct current.

Shown in Fig. 3-54 is an example of a circuit which uses the SCR as a standard rectifier of ac and as a switch. The input is house current which is rectified by the switched on SCR into the ac equivalent voltage in direct current. This is a half-wave rectifier circuit because only one-half of the ac sine wave is acted upon by the single SCR. The switching circuit is composed of a small resistor connected between the gate and the anode. When this resistor is switched into the circuit, a small amount of current flows between the gate and anode. The result of this action is that the SCR begins conducting current. Up until this time, the SCR appears as an open circuit and no current is delivered to the load.

The SCR will continue to conduct current as long as the switch is left in the on position. When the switch is turned off, the current flow will seem to cease immediately. For all practical purposes (in this circuit), it does. However, the SCR will not cease its conduction until the input current reaches zero. This occurs every 1/120th of a

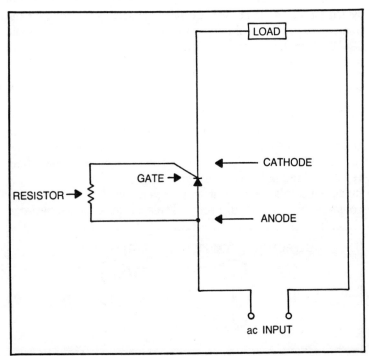

Fig. 3-54. A silicon controlled rectifier circuit.

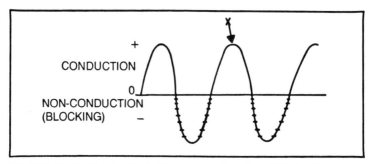

Fig. 3-55. Graph showing the half-wave operation of an SCR circuit.

second in a 60 Hz ac circuit, which is used as the source of electrical power in the circuit under discussion. Thus, if the switch were thrown off during the peak of one cycle which the rectifier was conducting, the device would not cease conduction until the cycle completed its decay to zero.

You can understand this further by referring to Fig. 3-55. Here, the ac sine wave can be seen as it would appear on an oscilloscope. This is a half-wave rectifier, which only conducts during one half of the cycle. In this case, it is conducting during the portion of the cycle indicated by the curves on the top portion of the zero line, or the positive cycle. During the other half of the cycle, it is blocking the flow of current. If the switch were to be thrown during the part of the cycled marked X, the SCR would not cease its conduction until the cycle peaked and returned to a zero value.

Although this 1/120th of a second lag may seem insignificant, it becomes quite important when a dc power source is used with the SCR being utilized for control and switching. Figure 3-56 shows the same basic circuit with dc substituted for ac. Since a rectifier will pass the flow of current in only one direction, it will easily conduct dc which is of a fixed polarity. This assumes that the diode is connected with correct polarity observation, as is done in this circuit. To reverse the diode would result in a zero current output.

As before, a small resistor and switch set up in a circuit between the gate and the anode. When the switch is thrown, the SCR will fire and current will be conducted through the load. When the current is switched off, there will be no change, because the SCR will continue to conduct until the current it is passing drops to near zero. The ac circuit did this several times in a fraction of a second, but dc stays at its constant value at all times of operation. The result is that in a circuit with a dc power supply, the SCR will continue to conduct regardless of the position of the gate/anode

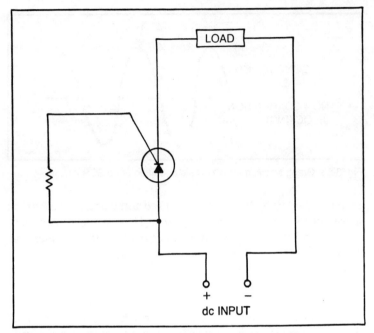

Fig. 3-56. The SCR circuit may be used with a pulsating dc drive.

switch until the source of current is removed from the circuit. At that time, the SCR will quickly return to its nonconducting state and it will be necessary to turn the gate/anode switch to the on position to return it to a conducting state again.

Triacs

A triac is another semiconductor device and operates in a somewhat different manner that the SCR already discussed. A triac, when triggered, will pass current in either direction. while triacs are really ac devices which control the flow of alternating current, they have applications in dc power supply circuits at the primary input to the power transformer.

It is already known that a SCR's output will always be direct current. This is suitable if the load needs or will operate from dc. However, if it is desirable to supply alternating current to a load, the SCR will not do. Figure 3-57 shows the use of a triac to provide just this. Here, two SCRs are combined in reverse parallel. The anode of one is connected to the cathode of the other and the cathode of the first is connected to the anode of the second. The gates are tied together. In this manner, current of one polarity will be conducted through one SCR; and when the current reverses, it will be conducted through the other. Both diodes are located within the same major circuit, so alternating current will be passed by this combination device when both are triggered. Fortunately, it is not necessary to use two SCRs to perform in this manner. The triac is a single, compact component which is about the size of a single SCR. The circuitry is connected internally through a chemical semiconductor manufacturing process.

Figure 3-58 shows a simple circuit that uses the triac to control current flow through an alternating current load. A small-wattage, high-value resistor is used as the triggering mechanism in combination with a small single-pole, single-throw switch. The triac appears as an open circuit when in its nonconducting state; but when the triggering switch is thrown, conduction begins. As with the SCR, the triac will continue to conduct until the current source reaches zero. Remember that in ac circuits, this action occurs every 1/120th of a second, so the switch may be turned off and the circuit will cease conduction within an instant. The switch must be left closed at all

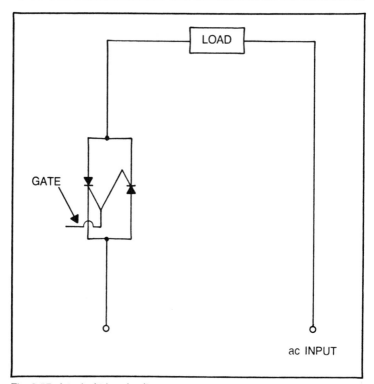

Fig. 3-57. A typical triac circuit.

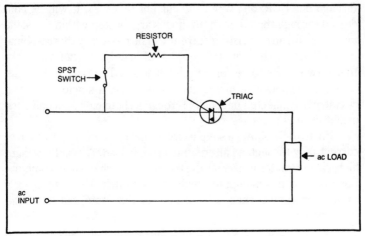

Fig. 3-58. A simple triac circuit for alternating current control.

times of operation for this reason, since if the switch is turned off, the conduction ceases almost immediately.

The triac is usually an expensive device although it is available in a configuration that will handle large amounts of current. The main thing to watch out for when using triacs is that they usually have specific voltage and current limitations. This is not a real danger, however, since if too much voltage is present, the triac will simply cease to conduct. Thus, the device itself will not be subject to damage. This is a distinct advantage over the SCR, since this device may be destroyed under the same conditions.

INTEGRATED CIRCUITS

An integrated circuit (often abbreviated IC) is a complete circuit which is integrated onto a single chip of semiconductor material, most often silicon. Integrated circuits see wide usage as complete regulator circuits in power supplies designed to produce very stable low-voltage outputs. They are also used in many different types of supplies which offer variable output voltage ratings.

The integrated circuit is composed of the same types of materials used to manufacture the discrete devices already discussed, such as transistors, diodes, etc. An integrated circuit may consist of these same devices all housed in one compact unit. The amazing thing about an integrated circuit is that although it may contain many transistors or other solid-state devices, resistors, and other components, it will usually only weigh a fraction of an ounce and can be held between the thumb and forefinger of your hand. An integrated circuit can contain over 1000 solid-state devices, and this tiny chip has had a revolutionizing effect on the entire electronics industry.

In an integrated circuit, all of the components used must be processed from the same solid-state material. The manufacturing process may differ somewhat from company to company, but the basic process remains very similar. The silicon or other material is treated with impurities to form a semiconductor. The various types of semiconductor materials are combined to form the different components. The chip of silicon used in an integrated circuit is simultaneously treated in a manner which causes microscopic portions of the water to be transformed into the different components. One tiny part then acts as a transistor, for example, while another will be a diode or a resistor.

An integrated circuit can be wired to form several different circuits, depending upon the wiring configurations used. Shown in Fig. 3-59 is an integrated circuit as it would appear if discrete components were used to form the same circuit. A circuit made with discrete components would be much larger than an integrated circuit. This enormous savings in physical space is responsible for the compact size of many types of electronic equipment available today. It is also important to note that one integrated circuit can be treated as a discrete component when it is combined with other parts of a larger circuit. The only difference is that instead of thinking in terms

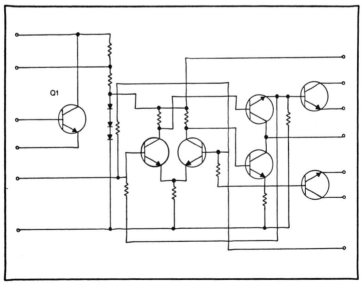

Fig. 3-59. A schematic drawing of the internal makeup of an integrated circuit.

of an individual component, we are now thinking in terms of completed circuits on tiny chips.

The integrated circuit is known for its dependability and ruggedness. By using integrated circuits without any external discrete devices, dependability can be increased even further. The greatest reliability is obtained when a single integrated circuit is used with a minimum of external connections or devices. ICs are impact-resistant, not subject to vibration damage, and can be enclosed in a compact case for protection.

The difference in the cost of an integrated circuit when compared with the cost of purchasing the same discrete devices to make up the same circuit is quite significant. An integrated circuit can cost as little as $2.00, while the discrete devices, if purchased individually, could be as much as $500. This makes the integrated circuit much more desirable for not only large manufacturers of electronic equipment, but for the home experimenter as well. It can be seen why the integrated circuit has quickly replaced many discrete components in instances where it is practical.

LIGHT-SENSITIVE SOLID-STATE DEVICES

Although solid-state devices have been available for many years, light-sensitive devices are not as familiar to some persons. Light-sensitive devices, as the name implies, respond directly to the presence of light. In other words, instead of responding to a source of current, these devices are triggered by a source of light, either natural or artificial.

Light-sensitive solid-state devices are finding applications in many electronic circuits today, including dc power supplies. They can serve to measure the intensity of light, as daylight warning switches, or even for communications purposes. Light-sensitive components can be divided into two major categories, depending upon the effect which occurs when they are exposed to a light source. These are photo-conductive and photo-emissive. An example of a photo-emissive device is the solar cell, which generates electrical current when light strikes its surface. In a photo-conductive device, the resistance to electron flow is changed by the amount of light which strikes its surface. The cadmium sulfide cell is a good example of a photo-conductive device. Those which generate electrical current in response to light can be thought of as active, while those which change their resistance to the flow of electrons are passive. There are many types of light-sensitive solid-state components available with a variety of applications.

SOLAR CELLS

The solar cell is a photo-emissive device and is, in itself, a wholly contained dc power supply. Whereas conventional power supplies draw power from the ac line, the solar cell gets its primary input from light. The solar cell is sometimes referred to as the photoelectric cell and can be directly utilized to power a circuit which requires low voltage and current. Although the shape may vary, a solar cell usually consists of a flat plate with a specially treated surface. This device is sometimes likened to a dry cell battery because it consists of a positive and negative terminal and can be connected in a parallel circuit for increased current or in a series circuit for higher voltage output.

Shown in Fig. 3-60 is the schematic representation of a solar cell. These devices normally have an output of approximately 0.45 volt. The output from the solar cell is dependent upon the amount of light which is present at its sensitive surface. Although the cell has a maximum output, greater light levels or intensities will create higher current ratings until this maximum is reached. The light energy is transformed into electrical current.

As might be expected, the solar cell has a quite high price tag, although in recent years, the price has dropped considerably. This reduction can be attributed to technological advances with regard to manufacture and efficiency. It is assumed that as further research information is made available, the price will become much more competitive, making the solar cell a viable alternative to its more standard counterparts.

SUMMARY

While many electronic devices can make up a dc power supply, in comparison with other types of electronic circuits, power supplies are quite simple and can be easily built from readily available parts. This chapter has discussed most of the components which will be found in the construction projects in this book.

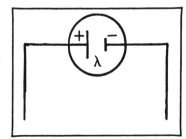

Fig. 3-60. Schematic representation of a solar cell.

It is not necessary to fully understand the exact technical operation of each electronic component, but it is important for you to have a general knowledge of each part's function in power supply circuits. Component ratings are most important and exceeding them can quickly lead to expensive failures. The schematic drawings in a following chapter devoted entirely to do-it-yourself projects will also include component lists which will provide the needed information to choose components which are correctly rated to function in these circuits.

Chapter 4

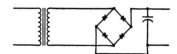

Voltage Regulators

Most electronic equipment performs satisfactorily with some variations in supply voltage. However, operation of many circuits is sensitive to minimal changes in supply voltage. Thus, use of a voltage regulator is required. In this chapter you will learn about the common types of voltage regulators.

An electronic voltage regulator is connected between the power supply and load impedance to maintain the output voltage at a specific value. The regulator circuit reacts automatically within its design limits to compensate for deviations in output voltage due to changes in line voltage or variations in load impedance.

Regulation of power supply output voltages using zener diodes is suitable for circuit loads requiring small currents, but these cannot regulate over a wide range. Current handling capabilities, as well as percent of regulation, can be improved by using transistors or electron tubes in conjunction with the zener diode.

Depending upon their circuit relationship to the load impedance, electronic voltage regulators may be grouped into two general types. These are shunt regulators and series regulators. Series regulators are the most efficient and therefore are the most widely used.

SHUNT REGULATOR

Figure 4-1 illustrates the equivalent resistive network of a shunt regulator. The locating of the regulating device is shown in

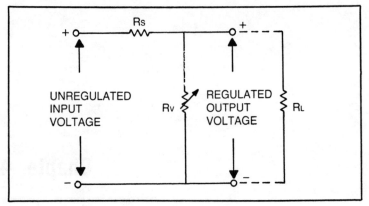

Fig. 4-1. Placement of a shunt regulator in relation to load impedance.

relation to the load impedance (R_L) Regulation is accomplished in the shunt regulator by division of current between the regulating device and load impedance. When the load demands more current, less current is diverted through the regulator (R_v), allowing more current flow through R_L.

Efficiency of shunt regulators is low under light load conditions (i.e., when current flow through the load is minimal), because the majority of current is being drawn by the regulator. Under full load conditions, regulator efficiency is high because minimum current is drawn by the regulator. A significant advantage of a shunt regulator, as opposed to a series regulator, is that it will not be overloaded under load short circuit conditions.

ZENER VOLTAGE REGULATOR

The breakdown diode, or zener, is an excellent source of variable resistance. Zener diodes come in voltage ratings ranging from 2.4 volts to 200 volts, with tolerances of 5, 10, and 20 percent and with power dissipation ratings as high as 50 watts.

The zener diode will regulate to its rated voltage with changes in load current or input voltage. Referring to the zener diode shunt-type regulator in Fig. 4-2, the zener diode VR1 is in series with fixed resistor R_S. The voltage across the zener is constant, thus holding the voltage across the parallel load R_L constant. Although the circuit shown depicts a positive voltage output, it is a simple matter to have a negative output voltage by reversing the zener and the polarities shown in this figure.

The value of R_S in this figure has been fixed so that it can handle the combined currents of the diode and the load and still allow the

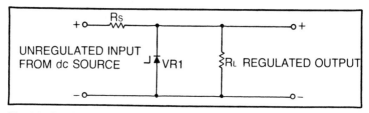

Fig. 4-2. Simple zener diode shunt regulator circuit.

diode to conduct well within the breakdown region. R_s stabilizes the load voltage by dropping the difference between the diode operating voltage and the unregulated input voltage.

The zener diode is a PN junction that has been specially doped during its manufacture so that when reverse biased, it will operate at a specific breakdown voltage level. It operates will within its rated tolerance over a considerable range of reverse current.

Again referring to Fig. 4-2, the operation of the zener diode shunt regulator will be discussed. If input voltage to the regulator circuit decreases, the voltage across the zener diode must also decrease, causing zener current to decrease. The total current in the circuit decreases, much the same as when the value of R_v increases in a simple shunt regulator, as shown in Fig. 4-3. The current through R_s, having decreased, results in a lower voltage drop across R_s. This results in the voltage drop across the zener and the load returning to the desired voltage. The zener diode makes the necessary adjustment automatically with the change in input voltage or load current.

When input voltage increases, the change is immediately felt across the zener. This effectively biases the zener so that there is an

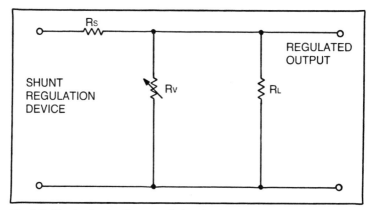

Fig. 4-3. Simple shunt voltage regulating circuit.

increase in zener current. The increase in zener current means an increase in total circuit current. R_s, in series with the source, will have an increase in current through it, resulting in a larger voltage drop across it. With the larger drop across R_s, the voltage across the zener and the load is reduced to the desired output voltage. The zener, in this instance, has a lower resistance.

For changes in load current, the zener makes the adjustment so that source current remains constant and the output voltage will also be constant. For instance, if the current drawn by the load decreases, the zener current will increase a corresponding amount. The total current remains the same. If the voltage across R_s is the same as it was before the load current decrease, if source voltage has not changed, then it is logical to infer that the output voltage is at the desired regulated amount. Conversely, if the current drawn by the load increases, the zener current decreases by the same amount and total current in the circuit remains the same. Since total current is the same, the voltage drops across R_s and VR1 (and therefore R_L also) cannot change because the source voltage has not changed.

Some words of caution about the shunt type voltage regulator are in order. Do not operate this circuit without a load. If a no-load condition exists, the zener must dissipate more power than usual (the load power as well as its normal power). If this condition occurs, it is quite possible that the maximum power rating or the maximum reverse current rating of the zener will be exceeded and damage to the zener will result.

If there is a failure in the voltage regulator just discussed, the following checks should help in locating the trouble:

■ Check to find out if there is a load on the regulator circuit. The lack of a load might indicate a damaged zener as the source of trouble.

■ Check the dc voltage measurements at the input to determine whether voltage is applied and whether it is within tolerance.

■ Since the operation of this regulator is based upon the voltage divider principle, measurements of the voltages across the output terminals and across R_s might be necessary to determine if the output voltage is within tolerance or if the drop across series resistor R_s is excessive. Be sure to observe correct voltage polarity when making these checks.

■ If the load is shorted or if R_s is open, a voltage measurement across R_s will indicate source voltage. In both cases, there will be no output. It will not be necessary to check the value of R_s. Disconnect R_s from the circuit when making the measurements.

■ If the zener diode becomes open, there will be no regulation and the output voltage should be higher than normal.

■ If the zener diode becomes shorted, there will be no output.

In general, if the output voltage is above normal, it is an indication of an open circuit in the shunt elements, either VR1 or the load, or an increase in impedance of these same elements. An output voltage that is below normal is an indication of an increased value of R_s, a low input voltage, or an excessive load current due to a decrease in load impedance.

SOLID-STATE SHUNT VOLTAGE REGULATOR

Figure 4-4 depicts a solid-state shunt regulator. Q1, in parallel with the load impedance, is an NPN transistor with the collector connected to the positive side of the voltage supply and the emitter connected to the negative side through CR1, a zener diode. CR1, when reverse biased to its breakdown voltage by R2, maintains a constant reference voltage at the emitter of Q1. The base voltage of Q1 is determined by the setting of potentiometer R1. This voltage is adjusted so that the base is positive with respect to the fixed emitter potential, forward biasing Q1 and causing it to conduct. The setting of R1 determines the amount of current through Q1.

The regulated output voltage is equal to the available supply voltage minus the drop across R_s, the series dropping resistor. The voltage drop across R_s is controlled by the amounts of current drawn by Q1. Thus, the setting of R1 will determine the value of the regulated output voltage.

Regulation is accomplished in the following manner. If for any reason the output voltage increases, the drop across R1 will increase. This will cause an increased positive potential at the base of Q1. Since Q1's emitter is at a fixed potential due to CR1, the more positive base will cause Q1 to conduct more. Current flow through

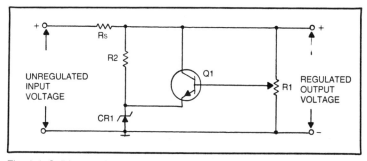

Fig. 4-4. Solid-state shunt voltage regulator.

R_s will increase due to the increased transistor current. This will cause an increased voltage drop across R_s, reducing the output voltage to the desired level.

For a decrease in output voltage, the regulation process is reversed. The decreased drop across R1 will decrease the forward bias of Q1, causing the transistor to conduct less. Current through R_s will decrease, causing voltage across R_s to decrease and increasing the output voltage.

The regulation process is essentially the same as that occurring in a shunt zener diode regulator. However, the current handling capabilities are greatly increased.

ELECTRON TUBE SHUNT VOLTAGE REGULATOR

An electron tube shunt voltage regulator is illustrated in Fig. 4-5. V1's cathode is maintained at a constant reference voltage due to the action of V2 and R2. The difference between the unregulated input voltage and the regulated output voltage is dropped across R_s. The output voltage level is determined by the setting of R1. V1's bias is the difference between the voltage at the wiper arm of R1 with respect to ground and the voltage at the plate of the VR tube with respect to ground. Just as in the solid-state version, changes in output are sensed by V1 through R1, varying the conduction of V1 and controlling the voltage drop across R_s in a manner that will maintain a relatively constant output voltage. This type of electron tube shunt regulator is used extensively in high voltage power supplies of video recorders.

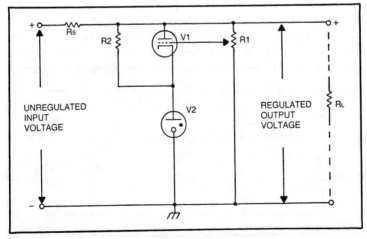

Fig. 4-5. Electron tube shunt voltage regulator.

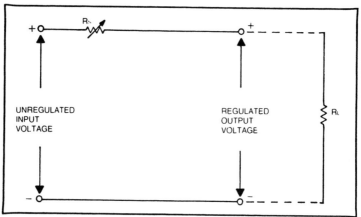

Fig. 4-6. Series regulator.

SERIES REGULATOR

Regulation is achieved in a series regulator (Fig. 4-6) by division of voltage between the regulator R_s and the load impedance (R_L), depending on the needs of the load. Efficiency of series regulators is high under light load conditions and low under full load conditions. The series regulator has no inherent overload protection. A short circuit in the load would cause heavy current through the regulator circuit, thus overloading most types.

Transistor and electron tube configurations of series, shunt, and shunt-detected series voltage regulators will be discussed individually.

SOLID-STATE SERIES VOLTAGE REGULATOR

A schematic diagram of a solid-state series voltage regulator is shown in Fig. 4-7. Q1 functions as a variable resistance between the source and load impedance. CR1, a zener diode, in conjunction with R1 maintains a constant voltage at the base of Q1.

The unregulated input voltage is applied across the series network of Q1 and R_L. The fixed base voltage is of sufficient value to forward bias Q1. Under normal operating conditions, the base bias is fixed at a value that will produce the desired voltage across the load impedance. The load voltage is equal to the unregulated input voltage minus the drop across Q1.

Regulation occurs in the following manner. Assume that the line voltage increases. This will cause a momentary increase in load voltage, which causes the emitter of Q1 to appear more negative with respect to ground, decreasing the forward bias of Q1. This

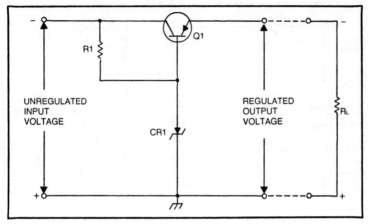

Fig. 4-7. Solid-state series voltage regulator.

decrease in forward bias causes the internal resistance of the transistor to increase, producing an increased drop across Q1 and thereby returning the voltage drop across R_L to its normal level. The action of the regulator is reversed with a decrease in applied line voltage.

Under varying load conditions, operation is a bit different. Assume an increase in load (more current). Voltage across the load impedance will decrease and the emitter of Q1 will become less negative, increasing the forward bias. This reduces the transistor's internal resistance, effectively reducing the voltage drop across Q1. This allows more current flow through R_L, thus returning the output voltage to its normal level. A decrease in load will cause a reverse reaction of the regulator.

ELECTRON TUBE SERIES VOLTAGE REGULATOR

An electron tube version of a series regulator is illustrated in Fig. 4-8. The operation is similar to a solid-state series regulator. V1 acts as a variable resistance between the source and the load impedance to compensate for changes in line and load voltage. V2 maintains the grid of V1 at a fixed reference voltage. If the output voltage increases, the cathode potential goes more positive with respect to ground, thereby increasing the bias. This increase in bias causes V1 to decrease conduction, decreasing the drop across R_L to the desired level. The opposite action will occur for a decrease in output voltage.

A disadvantage of simple series regulators is that they do not rapidly respond to small changes in voltage. The effectiveness of a series regulator is improved by the addition of circuitry that detects and amplifies small changes, thus allowing the regulator to respond more rapidly. The shunt detected series regulator is such a circuit.

SHUNT DETECTED SERIES VOLTAGE REGULATOR

Figure 4-9 depicts a block diagram of a shunt detected series voltage regulator. The series regulator acts as a variable resistance in series with the supply voltage and load impedance to compensate for any changes in source or load voltage. A portion of the output voltage is fed back to the amplifier and detector stage, which compares the sampled voltage with a previously set reference voltage and senses any change. This change is then amplified and sent to the series regulator, which changes the conduction level, thereby returning the regulated voltage to the desired level.

SOLID-STATE SHUNT-DETECTED SERIES VOLTAGE REGULATOR

A solid-state shunt-detected series voltage regulator is illustrated in Fig. 4-10. This circuit consists of two sections: the regulator circuit and the control circuit. Q1 acts as a series regulator and operates in the same manner as the circuit in Fig. 4-7. The control circuit is composed of Q2 and its associated circuitry.

The emitter of Q2 is held at a constant negative potential with respect to point A due to the action of R1 and CR1. Resistors R3, R5 and potentiometer R4 act as the voltage divider network in parallel with R_L. A negative potential with respect to point A is tapped from R4 and applied to the base of Q2, controlling its forward bias. R2 establishes the forward bias level for Q1. Thus, the intensity of current flow through Q2 determines the ultimate conduction level of Q1, either manually or automatically.

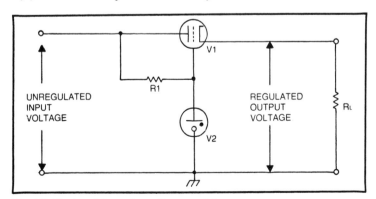

Fig. 4-8. Electron tube series voltage regulator.

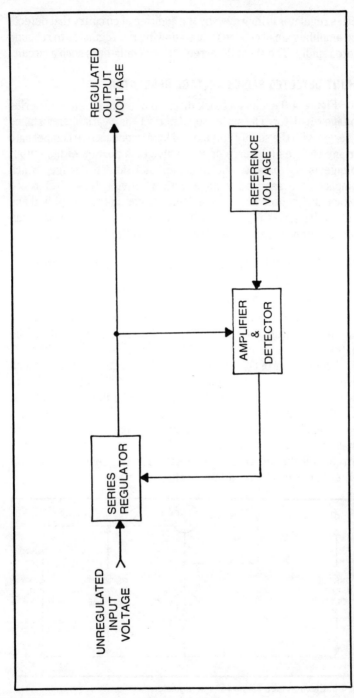

Fig. 4-9. Block diagram of a shunt detected series voltage regulator.

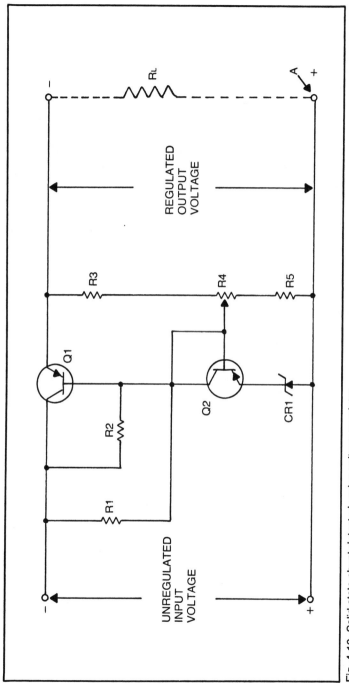

Fig. 4-10. Solid-state shunt detected series voltage regulator.

R4 is initially adjusted to fix a conduction level of Q1 which establishes the desired load voltage. If load voltage increases due to either an increase in source voltage or a decrease in load current, the negative potential with respect to point A at the wiper arm of R4 will increase, increasing the forward bias of Q2. This increase in forward bias causes Q2's conduction level to increase and collector voltage to decrease, reducing the negative potential at the base of Q1 with respect to point A. This decrease in negative potential causes the forward bias on Q1 to drop, decreasing current flow through the series network of Q1 and R_L and returning the output voltage to its preselected level. A decrease in load voltage will produce the opposite effects as outlined here.

ELECTRON TUBE SHUNT-DETECTED SERIES VOLTAGE REGULATOR

A schematic diagram of an electron tube shunt-detected series voltage regulator is shown in Fig. 4-11. V1 performs as a series regulator and the control circuit consists of V2 and its associated circuitry. The plate of V3, a VR tube, is connected to $+E_{bb}$ through R2. This allows the VR tube to ionize, keeping the cathode of V2 at a fixed potential with respect to ground when the input voltage is applied. A positive voltage with respect to ground is tapped from R4 and applied to the grid of V2, controlling the tube's bias. R1, plate load resistor for V2, establishes the grid bias level of V1. Since V2's plate is directly connected to the grid of V1, the amount of current flow through V2 determines the conduction level of V1.

Once R4 is initially set, any change occurring in source voltage or load voltage will cause a change in the potential across R4, changing the bias of V2. Assume a decrease in the source voltage. This will cause a momentary decrease in output voltage and reduce the potential at the wiper arm of R4, increasing the bias of V2. This increase in bias will cause V2 to reduce conduction, decreasing the drop across R1. The decreased drop across R1 causes bias on V1 to decrease, reducing its internal impedance. With less impedance, the voltage drop across V1 decreases, counteracting the decrease in load voltage. An increase in load voltage would cause the opposite effects throughout the circuit.

A pentode is used in the control circuit to take advantage of its high amplification factor. Even rapid variations in load voltage will be amplified sufficiently by V2 to change the conduction level of V1. A pentode will give a better percentage of regulation and a faster response than a triode. There are many variations and modifications

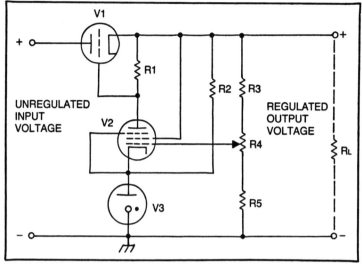

Fig. 4-11. Electron tube shunt detected series voltage regulator.

of the basic shunt detected series voltage regulator. A few examples will be discussed in the following paragraphs.

Since total load current flows through the series regulator, this device must be capable of passing a high value of current. Several transistors or electron tubes may be connected in parallel if the current capability of a single device is not sufficient. This configuration is illustrated in Fig. 4-12.

Another modification is to insert an additional dc amplifier, as shown in Fig. 4-13. The amplifier detector stage performs the same function as in the basic configuration, but now the output is fed to the second dc amplifier. This increased amplification provides greater sensitivity, thereby improving percent of regulation and speed of response.

If a stable voltage source is available, such as another regulated supply, the zener diode or VR tube can be eliminated. The reference voltage would then be provided by the constant voltage source.

CURRENT REGULATOR

Some types of equipment require a stable current supply. Figure 4-14 illustrates a simple solid-state series current regulator. This circuit will supply a constant value of current under varying conditions. Its operation is similar to a simple series voltage regulator.

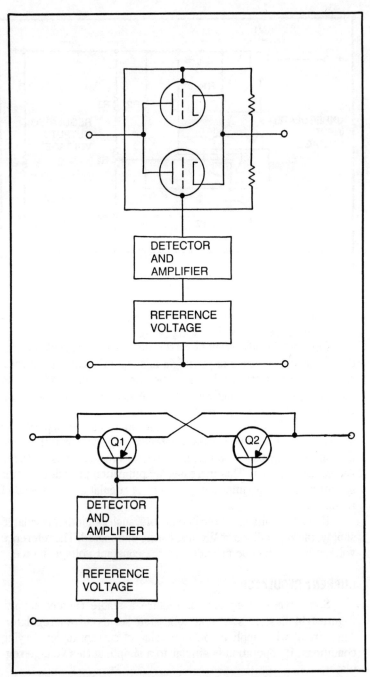

Fig. 4-12. Parallel shunt detected series voltage regulators.

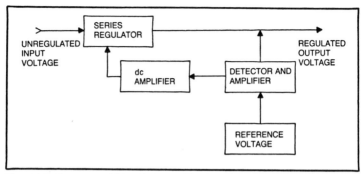

Fig. 4-13. Here, an additional dc amplifier has been inserted for increased amplification.

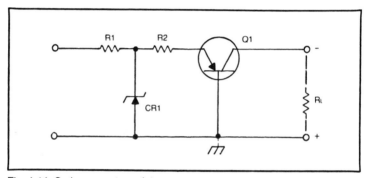

Fig. 4-14. Series current regulator.

Regulation occurs in the following manner. If current increases, the voltage across R2 will increase with respect to ground. This increase will reduce the forward bias of Q1, increasing the transistor's internal impedance and thus reducing current flow to the predetermined level.

Chapter 5

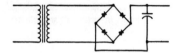

Obtaining and Referencing Components

Many would-be electronic enthusiasts shy away from building because they feel they have no sources for obtaining many of the electronic parts needed. There is absolutely no reason for anyone interested in building electronic projects to feel this way, especially when referring to the construction projects in this book. True, some projects you may see from time to time will call for an unusual part or two. Often, however, a source of supply is stated in the project description. In this book, none of the components used for completing the circuits are highly unusual, but some which are not as common as others will reference suppliers who carry them. Many parts can be directly replaced by equivalents from other manufacturers.

The place to start in being able to locate electronic components and accessories is at your local hobby stores. Here, you will find 90 percent of the items used to build the great majority of circuits in this book. These same stores will most likely carry mail order catalogues from such firms as Allied Electronics, Fair Radio Sales, and others that have been suppliers of electronic components for many years. The latter supplier sells mostly government surplus materials at very reasonable prices, while Allied Electronics is a major supplier of new components and devices ranging from transistors, diodes, photocells, etc., to electronic computers and industrial electronic equipment. These are just two mail order suppliers. There are many, many others.

A trip to a local electronics repair facility can quickly get you the names of parts suppliers in your area. You may even be able to order any electronic part imaginable through these sources. Once you have the name of a parts supplier, a phone call should get you a complete assortment of catalogues from the manufacturers whose projects they handle. Typically, you will find catalogues from electronic component manufacturers such as RCA, Motorola, Sprague, Sylvania, Miller, Amphenol, General Electric, and many others. Just the ones mentioned here will probably be adequate to supply 99 percent of your needs. If this supplier is located in your town or city, this will probably mean same-day service if the parts you require are in stock. If you live in a rural area, chances are these suppliers have route men who pass through on a regular basis. A phone call to the supplier will usually mean that your order can be delivered within a couple of days. If you elect to go the mail order route, a phone call to their order desk (often a toll-free 800 number) will allow your order to be mailed out the same or following day.

The cost factor may drive a lot of experimenters away from electronic project building, and there is a great fluctuation in prices of equivalent components from various suppliers and dealers. For this reason, it is absolutely essential to shop around and find the best deal. The author has economically replaced solid-state components called for in a circuit with a type from another manufacturer. The two parts were electronically equivalent; however, one was less than half the price of the other. This is where a lot of unnecessary expense may enter the electronic building hobby. This especially applies to solid-state devices such as diodes, integrated circuits and transistors.

CROSS-REFERENCING

One of the most valuable aids which can be obtained from the various suppliers is cross-reference material for the manufacturers who make solid-state devices. Each manufacturer prints its own catalog or replacement guide which lists almost every transistor, diode and integrated circuit made today and tells you which devices from this particular manufacturer are direct replacements for them. It will be necessary, or at least highly desirable, to have replacement guides from as many manufacturers as possible. For this discussion, three solid-state component manufacturers, Sylvania, General Electric, and RCA, have been chosen to demonstrate device cross-referencing. Let us assume that an electronic circuit calls for a 2N403 transistor. The 2N designation is one that is rarely used anymore. It can be considered as the old generic name for transistor devices. Today, most manufacturers have an individual and specialized numbering system for naming their devices. In search of a 2N403 transistor, we first look through the Sylvania replacement guide. It reads just like a progression list. First we locate the prefix 2N and work from there. Reading up the list, we have 2N399, 2N400, 2N401, 2N402, and finally, 2N403. To the right of the 2N403 reference is the Sylvania replacement. We find that an ECG 102 is a direct replacement for the 2N403. The same can be done using an RCA catalog, which will tell us that the SK3003 is a direct replacement for the 2N403. It is also a direct replacement for the ECG 102. The General Electric replacement guide specifies a GE-53 as the exact replacement for any of the devices mentioned so far. If the transistor number supplied in the circuit diagram were that of another manufacturer, this device could also be referenced to the current device equivalent from another manufacturer. Assume that a transistor is called for which is a Sylvania ECG 107. The RCA replacement guide would list the ECG 107 along with the RCA equivalent, which is an SK3203; or the General Electric guide would specify a GE-11. All of these devices are electronically equivalent. The great majority of them will contain the same case and lead configurations; however, it is often good to make certain that the leads are arranged in identical order. Sometimes, different cases will be used to house an equivalent crystalline chip, but these occurrences are usually specified in the cross reference manuals.

The vast numbers of transistor identifications often confuse and frustrate beginning and experienced experimenters. If you keep the cross reference manuals on hand at all times, you may find that you have an equivalent device already on hand, either new or contained in a junk box circuit.

THE EXPERIMENTER'S JUNK BOX

Every able experimenter accumulates a "junk box" of parts in a very short period of time. The term junk box is a misnomer, as many of the components can be put to good use in future projects. A junk box may actually be several crates, boxes, and cabinets filled with surplus transmitters, receivers, old data circuit boards, electric clocks, etc. Through many surplus outlets, it's possible to obtain brand new circuit boards from discontinued lines. Manufacturers dump a lot of outdated components and circuits into these industrial surplus channels, and some very good buys can be made. It is not uncommon to purchase a transistor or IC from one of these mail order dealers which is guaranteed to be new and operational and which sells for one-tenth of the price of an equivalent component on the retail market. Rectifier diodes may be purchased for 11 cents apiece which might cost $2.50 if purchased from a retail repair facility. Recently, when several citizens band radio manufacturers went bankrupt, huge numbers of their completely wired circuit boards were available through surplus channels. They sold for less than $10 and contained close to $100 worth of brand new parts. With a soldering iron and some desoldering tape, most of these components can be salvaged in like-new condition. The leads may be a little short, but the low price is worth the extra difficulty encountered in their salvage.

Enterprising experimenters keep track of these products through ads placed by the distributors and dealers in electronics magazines. They also send a letter to the distributors asking that they be placed on a regular mailing list to receive constant updates on the products that are available. When a good buy comes along, one which offers components of immediate or future use, a purchase is made. These parts are stored away until needed. This is what constitutes a junk box. Sometimes, a buy is made which was so attractive due to low price without any idea of what the purchase may be used for. Intuition plays a role here, and occasionally these purchases turn out for the best. Often, however, the components bought sit around for a few years and are traded off to other experimenters, resold, or tossed in the real junk box, often called the trash can. In the early stages of building up a parts supply, it is very easy to become enamored with having a lot of *anything* electronic, regardless of whether or not it would seem to have a future use. This is an excellent way to waste money. Don't become so parts-conscious that you must have everything that is available at a ridiculously low price. Some components are this cheap because their practical uses are ridiculously few and far between.

There is another aspect to building a junk box which costs little or nothing. Many amateur radio operators, CB enthusiasts, and experienced experimenters have junk boxes which have been transformed into junk rooms. Their amount of junk has reached a state where it is overflowing the owner's ability to keep it in an orderly fashion. In these situations, it may be possible to haul away wheelbarrow loads of components for free. This is more preferable to the owner than having to pay to have it hauled to the junkyard. Don't neglect your local radio and television stations either, nor nearby television and stereo repair shops. One might be surprised at the wealth of electronic components which may be obtained from an old television receiver. True, it is a tedious and often dirty job to remove all of the components, but often worth the effort.

The author has been successful in salvaging many electronic parts from printed circuit boards in a rapid fashion by cutting the circuit board out from under its components. The only tools required are diagonal cutters and needlenose pliers. Most of the older types of circuit board material will chip away in large pieces when cut. The green, glass-epoxy boards will present major problems with small cutters, and a larger tool might be brought into play. Usually, the entire circuit board can be cut from the components without having to use a soldering iron. Gobs of solder which remain at the ends of the component leads turn to dust when tightly squeezed in the jaws of the long-nosed pliers. Medium-sized circuit boards with 50 to 100 components can be cleared within a half hour to forty five minutes. True, a few components do get damaged from this rather hasty method, but any especially critical devices may be removed ahead of time by desoldering the leads in a conventional manner. Clip-on heatsinks can be used to protect them if necessary.

Your local electronics hobby store may also be a source of almost free junk box parts. The local Radio Shack store in the author's city often displays defective or damaged merchandise on a special counter which can be had for pennies on the dollar of retail cost. Additionally, discontinued items may also wind up here at reductions of over 75 percent. Many of the damaged items can be repaired, while others can be salvaged for whole circuit sections or individual components.

Don't neglect the mechanical parts either. These will be found through some electronic outlets and in the junk boxes of other experimenters. These include switches, relays, and mechanical timers which may often be put to good use in various electronic circuits. Buying these parts new could cost $50.00 or more. This is ridiculous considering the fact that a surplus component might suffice just as well. It might even be better. One has to realize that government and industrial surplus parts may have originally cost hundreds or even thousands of dollars and may sell for less than ten dollars. The price differential quoted is not a highly unusual one in the world of surplus components and will be verified quite often as you continue to experiment and obtain parts.

The previous discussion has dealt with obtaining electronic parts in general. Power supplies will utilize many of these components, but in one area, you may not be able to find just what you want on the commercial market unless you go to a great deal of expense. This involves power transformers, especially those designed for medium and high voltage output. Your local electronics store will most likely be able to supply you with inexpensive low voltage transformers whose secondaries are wound to produce anywhere from a 6 to 25 volt output. These usually sell for less than ten dollars. But what if you want a transformer with a 500 to 800 Vac output? You will probably have to order these through a commercial outlet and they can cost well over fifty dollars. This is a little steep for the experimenter who may wish to build several different medium voltage supplies.

Fortunately, there is still a ready source for these types of transformers; often they can be had for absolutely nothing. In my opinion, there is nothing quite so valuable to the power supply experimenter as a junk black and white television receiver. The older types, made in the fifties and sixties, will be especially useful. Transformers in these sets usually have several secondary windings. Typically, one will be rated at 5 Vac, another at 6 or 12 Vac, and the third will produce an output of between 500 and 800 Vac. Power ratings will vary, but these are generally heavy duty units rated at 200 to 300 watts, even higher when used for ICAS duty.

The more modern color television receivers contain heavy duty transformers as well, but many of these have high voltage secondary windings of less than 300 Vac. These can certainly be used for many applications but may not be as valuable when a final dc output from the power supplies in which they are used is to fall into the medium or high voltage range.

Television repair outlets usually haul several sets per week to the local junkyard and are more than happy to have some experimenter truck the whole lot away free of charge. It is usually a simple matter to remove the backs from these sets and start gathering components. The first thing to do is remove the chassis. The cabinet and picture tube can then be taken to the junkyard. Locate the primary and secondary leads from the power transformer and snip them at their contact points. These should take only a few minutes. Then, with a screwdriver and nutdriver, the four mounting bolts which hold the transformer case to the chassis are removed. The transformer will now slide free.

Don't stop here, as you will undoubtedly be able to salvage quite a few resistors and maybe even a filter capacitor or two. There will also probably be a good assortment of switches and other mechanical components which can be put to direct use in a homemade dc power supply. You may even be able to salvage a solid-state rectifier or two, although it is sometimes easier to simply purchase these low-cost components from an electronics outlet.

Once the power transformer has been removed, it is necessary to check it for proper operation and also to identify the various leads. A voltmeter which will measure ac voltage up to about 1,000 Vac is necessary. First of all, find the primary leads. These are usually the ones which contain a solid black insulation, but often age will turn all of the leads to a near black color. During the transformer removal procedure, the primary leads are more easily identified as they connect to the ac line cord through a fuse block. Also, the primary leads are usually located on one side of the transformer, whereas the secondary side will have four or more leads.

You must be very careful when measuring voltage at the secondary of a power transformer. Lethal potentials often exist here. Connect a line cord with plug attached to the primary transformer leads and make sure the secondary leads are not allowed to short together. Insert the plug into the wall outlet and with the voltmeter set to the 1,000 Vac range, begin taking measurements. Secondary lead insulation coloring will vary, but often a pair of green leads will be seen. This color normally indicates a filament winding for rectifier tubes. The measured value here will probably be 5 Vac. You may also see a pair of red leads which indicate the high voltage winding. A multi-colored lead is probably the center tap of this latter winding. To check this out, measure the voltage between the two red leads and then take another measurement between one of the red leads and the multi-colored one. If you get a reading which is half of the value which was obtained between the two red leads, then this identifies the center tap winding. Remaining leads will most likely carry a potential of 6 or 12 Vac.

It may take a little while to correctly identify the leads of each winding, but patience will pay off here. Be sure to permanently mark each lead in order to identify it for future use. This can be done by wrapping each conductor with paper tape which can be easily marked with the voltage value. Alternately, you can jot down the lead coloration and its voltage rating on a separate piece of paper.

As far as power ratings are concerned, you will probably have to make a good guess. The smaller models will probably be good for about 200 watts in intermittent service, while larger transformers will safely withstand a current drain of about 400 watts in the same type of service. Several years ago, the author removed a rather small transformer from an old black and white set. It was quite unusual in that fins protruded from the case which served as a heatsink. This unit was incorporated in a homebuilt ham radio transmitter and delivered a 300 watt output for many years. The transformer operated at a rather high temperature, but it was still functioning when the transmitter was sold. Amateur radio work involves ICAS service, so while the transformer was probably being operated outside of its manufacturer's ratings, the long cooling off periods between transmission probably prevented failure.

If the salvaged transformer is not rated to produce the amount of power output needed, you can probably combine two similar units in parallel to arrive at the output required. Also, two similar transformers may be used with their secondaries in series to produce a high total secondary value. When this is done, their primaries may be wired in parallel for 115 volt operation or in series when they are to be driven by a 230 Vac source.

The extra windings at the secondary come in handy too. For instance, the 5 volt filament source can be used to drive a rectifier circuit which will produce a 6 Vdc output. This should be good for a total drain of at least 1 ampere. Voltage doubler circuits can also be used with this or other windings to produce a wide range of dc output options.

War surplus equipment was previously mentioned and while you have to be careful about what you purchase, here is another excellent source for power transformers, filter capacitors, switches, etc. The author was fortunate enough to obtain a power supply circuit through one of these channels for less than $25. He didn't really know what he was buying at the time, but upon opening the chassis, he found a magnificent power transformer which delivered a multitude of output voltages at the secondary and had a primary winding that would operate at any potential from 90 Vac to 270 Vac. The transformer weighed over sixty pounds and was designed for a constant power drain of over 1,000 watts. The secondary windings consisted of outputs which ranged from a high of 2,500 Vac to a low of 2.5 Vac. To top all this off, the unit was brand new and the transformer was completely enclosed in a weatherproof case. This supply obviously was designed for the Navy and had been protected against salt air and moisture.

This purchase also contained a large number of high voltage fuses, heavy-duty switches, several relays, and even a small blower for cooling purposes. Again, the transformer had never been used and was in absolutely perfect condition. If this transformer were available on the commercial market today, it would probably sell for well over $300. The additional parts obtained would have easily run another $100 or so if purchased through normal channels. This was an excellent buy and is indicative of some of the bargains which can be had through surplus channels.

The thing to remember is that all types of surplus and junk equipment are good sources for power transformers. The medium and high voltage power supply enthusiast will especially want to look for tube type equipment. Here, transformers are often found with multiple secondary windings that offer many output values. Old PA systems, stereo systems, console radios, etc. can save the power supply builder well over $100 in components that would have to be purchased new otherwise. True, a defective transformer will be found occasionally, but more often, the equipment was relegated to junk status because of other internal problems. The author has procured several hundred pieces of junk equipment and only in one case was a power supply found to be defective. Often, the pre-punched chassis found in these types of equipment can be made to serve as a mounting base for the entire dc power supply. You may even be able to use the original cabinet.

If you wish to build a high voltage power supply with an output of 2,500 Vdc or more, there is another source which you may be highly interested in. Your local electric company may have a whole yard full of transformers which have been removed from service for various reasons. Often, these are large units which enclose the actual transformer in a case which is filled with cooling oil. When these cases begin to develop serious leaks, they are often removed from service, but the internal transformer is still good. All you need to do is remove the transformer from the case and use it as is. The cooling oil removes heat from the transformer and allows it to be used at very high power levels (typically 5 kilowatts or more). Without this oil, the transformer should still be good for well over 1 kilowatt. *Caution*: Recently, it was disclosed that a chemical known as PCB was present in some types of transformer oil. This is a known carcinogen. The power company has been advised to get rid of this type of oil, but some of it may still be present in old, discarded units. If you arrange a purchase with your local company, be sure to ask about this possibility. You may find units which have already been drained that sell for less than $25.

The transformers under discussion are actually step-down types. They are designed with a primary winding rated at 2,000 to 5,000 Vac. Their secondaries deliver 115/230 Vac, depending upon how they are connected. You previously learned that transformers are two-way devices. Therefore, your use of these units will involve incorporating the original secondaries as primaries and vice versa. By applying 115 or 230 Vac to the appropriate windings, you will get an output of 2,000 to 5,000 Vac, depending upon the type of transformer.

If you are interested in going this route, check with your local power company and find out what the availability of damaged components is. Chances are, you can leave your name and phone number so that you may be called and advised of future availabilities. These units are especially good buys because they can easily cost close to $1,000 or even more if purchased new. We are talking now about uncooled equivalents of what is available from the power companies. Oil-encased transformers designed for commercial service normally cost in the many thousands of dollars. When doing your research work, be sure to specify a single-phase transformer. Undoubtedly, some three-phase models will have also been junked and are not appropriate for most applications.

Regardless of what type of transformer you need, a thorough search will often turn up several possibilities which will cost little or nothing. Resourcefulness is the key here and may save you many hundreds of dollars.

KEEPING TRACK OF ELECTRONIC COMPONENTS

Once you have built up an adequate supply of electronic components, the problem emerges as to how to keep track of everything you have. The reason for stocking up in the first place was to allow you more flexibility in being able to tackle simple electronic projects without having to order every part required. But if you can't lay your hands on what you've collected or are not sure of all that you have, the original purpose for obtaining the parts is defeated.

Fortunately, it is very simple and inexpensive to categorize the great majority of your surplus components, and the entire process need take no more than a leisurely weekend. To repeat, parts categorization is absolutely essential to the operation of a successful experimenter's workbench.

First of all, begin with the heavy components. Large transformers, chokes, and similar devices may all be placed in a wooden or cardboard box and stored in a closed cabinet or on a shelf. The side of the box should be plainly marked to identify the materials it contains. If you have a great deal of different transformers, perhaps all of the low voltage units could be in one box, while the medium and high voltage parts are placed in two others. Each would be labeled accordingly. Cigar boxes make excellent storage trays for the components which are small enough to be contained by them. Small audio transformers, switches, relays, etc. may be kept here. The front or side of the cigar box is marked with a black pen to identify the contents. While it is a good idea to house only one type of component in each box, it is also possible and practical to keep a variety of components which are interrelated as to their uses in circuits. For example, one box might be marked AM Radio Parts and might include capacitors, ferrite rod antennas, speakers, etc. that were removed from AM radios. Cigar boxes can be stacked one above the other and contained in a relatively small space. They are very sturdy and will last for many years if stored away from damp areas.

For the individual components such as resistors, capacitors, and solid-state devices, you might purchase workbench shelves with plastic racks. These are available from discount stores and many other sources. Shown in Fig. 5-1, a typical cabinet may contain twelve to over forty plastic trays which slide forward to give access to their contents. Transistors may be categorized and placed in each drawer. Often, these drawers can be subdivided into three different compartments using plastic spacers contained with each purchase. Three different kinds of transistors, diodes, capacitors, etc. can be stored in this single drawer and kept separate from one another. Figure 5-2 shows a small parts cabinet which is especially designed for stacking purposes. This is more apparent when looking at the top

Fig. 5-1. The trays on this cabinet slide forward to provide access to components.

Fig. 5-2. A typical small parts cabinet which can be stacked.

and bottom of the unit. The bottom is recessed and will slip over the raised, mating surface which is found at the top of a similar unit. Figure 5-3 shows how several units may be stacked to conserve space on a workbench.

Resistors should be stored according to ohmic value. Figure 5-4 shows a resistor storage cabinet which is commercially manufactured and contains several different compartments in each drawer. Each of these compartments has been marked with a value. Fortunately, ohmic values of carbon resistors are pretty much standardized. For example, there is a 2700 ohm and 3000 ohm resistor, but none are commonly manufactured for 2800 ohms. Since resistance values are standardized, it is quite simple to purchase a resistor cabinet which is marked to identify common carbon resistors. Of course, if you prefer, parts cabinets similar to the ones

Fig. 5-3. Several units may be stacked to save space.

Fig. 5-4. A typical resistor storage cabinet with compartments labeled with various values.

previously discussed can be marked with resistance values and used in place of a commercially manufactured cabinet.

It is also a good idea to equip your workbench with several large plastic trays which can be used to house various "pull-out" components until they can be properly categorized and stored. Always discard damaged components or those which are never to be used. These have a tendency to build up and become intermixed with the categorized components. The job of keeping an orderly parts system is mainly comprised of throwing away what you don't need, rather than keeping what you've already filled in a constant state of order. If you do the former, the latter will usually take care of itself.

If you're going to keep dry cell batteries, keep them in metal drawers or boxes. Make sure these compartments are leak-proof. Check battery storage trays periodically in order to identify and discard any cells which may be leaking. These can quickly damage the remaining good cells and a leak can ruin other components and finished circuits.

This storage system should encompass the majority of the parts and components you are likely to collect. There will always be some devices and circuit parts which cannot be accurately filed away, and these must be relegated to a *Miscellaneous* box or cabinet. Whenever possible, the materials filed under this category should be placed in a more appropriate area as time permits.

With this type of filing system, it should be much easier to lay your hands on the parts which are needed to build a proposed electronic project. Once you know what you have, it is an easy matter through process of elimination to determine which components you will have to purchase. Once these are on order, your project is well on the way to getting off to a good start.

SUMMARY

Electronic components comprise many millions of different parts. Many of them are electrically or electronically identical but are given different manufacturer's numbers and designations. It is essential to have an understanding of these components and what they are expected to do in the circuits they are a part of. In the construction of solid-state circuits, it is important to obtain as many cross-referencing aids as possible. This will allow you to select components which are easily obtainable and as inexpensive as possible. When building circuits from older schematic drawings, many of the solid-state parts will not be sold by the manufacturer under the same designations which were given years ago. Fortunately, most of these parts are not obsolete. They will have modern equivalents which are identical to or better than the originals. It should be noted that manufacturer's equivalent parts may not always be exactly identical to those of other manufacturers. In most circuits, this will make little difference; but in a few, problems may occur. In cases such as these, you might wish to try another manufacturer or make circuit changes which might facilitate the use of another component type.

The circuits included in this book all use easily obtainable and currently available components. Sylvania solid-state products are specified most often in the chapter on electronic projects, because the author normally uses these in his experimental circuits. But, the reader may have another source which is more convenient. In this case, the Sylvania components can easily be cross referenced to any of the other major manufacturers such as RCA, General Electric, National Semiconductor, etc.

Many beginning builders are hesitant to substitute one component for another. It's good to be cautious in this area, but most substitutions can be experimentally tried without fear of circuit damage, providing that the builder uses a bit of common sense coupled with some manufacturer's specifications. For example, many of the circuits using transistors in this book will work with hundreds of different types, even though most of these will not be listed in the cross reference guides as exact replacements for the ones specified in the project schematic. The purpose of this book and the projects presented is for the builder to have fun, gain some learning experience, and to finish with a completed circuit which has a practical use or uses. Don't be afraid to experiment. That's what this book is all about.

Chapter 6

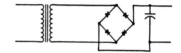

32 Electronic Power Supply Projects

Now that a groundwork has been laid in power supply basics and in the home building of electronic projects, it's time to move on to the actual projects themselves. All 32 of the electronic circuits presented in this portion of the book are of non-critical construction, use commonly available components and are relatively inexpensive to complete. Each finished project can be used to power other types of electronic equipment, and each circuit should provide a voltage output and current rating which is very close to design specifications if the stated components are used.

It is important that you think safety constantly throughout the construction and testing stages and then afterwards during the operational period. While many of these projects will produce a relatively low voltage output when completed and working, most of them derive power from the 115 volt ac line. As you already know, house current is a potentially lethal source of electricity, so caution is necessary at all times.

Don't hesitate to substitute components when it is necessary to build a power supply which is to have an outlet voltage of a different value than that specified in any of the projects. Remember, if you increase the transformer output voltage at the secondary, it will also be necessary, in most instances, to increase the voltage ratings of all components which lie past this point. Likewise, if a transformer is used which contains a higher current rating than the one specified in the schematic drawing, the values and ratings of components may need to be beefed up if this extra current is to be drawn from the supply. Naturally, if a specific schematic calls for a 1-watt resistor (for instance), a 2-watt resistor will serve just as well. But don't take the chance on using a ½-watt component, as it may fail. If a tranformer current rating of 1 ampere is specified, a 2-ampere transformer may be directly substituted with no other changes as long as the original 1 ampere of current is the maximum drawn from the finished supply.

Most of the line-driven transformers used in the following projects operate from the 115-volt ac line. If it is more desirable to use 230 volts as a source, it should be fairly easy to obtain transformers with 230-volt primaries and secondary windings which are equivalent to the secondaries in these projects.

All of the projects presented in this chapter are designed with flexibility in mind. No specific information is provided on cases or housings for these circuits, as many will be incorporated directly in other electronic equipment. However, any power supply which receives power from the ac line should and must be completely contained within an insulated or grounded chassis or cabinet. All precautions should be taken to absolutely prevent accidental contact between voltage points in these power supply circuits and interested passers-by. Also, all power supplies which operate from the ac line or which are connected to high current primary sources must be fused to prevent a fire hazard. Elimination of this most important part of the circuit can cause component damage should a malfunction occur, as well as overheating and a possible fire.

Projects 1, 2, and 3 are similar in that the output voltage and current for each project are identical to the other two. The methods used to arrive at the final output mark the distinction between each of the three. Of course, the regulation characteristics of the full-wave circuit make Projects 2 and 3 more suitable for a wide range of electronic uses. On the other hand, the simplicity and inexpensive nature of the half-wave circuit of Project 1 make it ideal for low current, non-critical applications or where a power supply must be "slapped" together in a few minutes.

While each of these first three projects is designed to deliver 12 to 15 volts under heavy to moderate loading, the three circuit diagrams can also be used to produce power supplies with totally different output voltages and current capabilities. All that is necessary is to alter a few component values, and you're there. In most cases, the value of the filter capacitor can remain the same (500 μF, but you should increase its working voltage rating. Naturally, the peak inverse voltage rating of the diode or diodes will need to reflect the ac voltage output value of the transformer secondary. Silicon diodes with 1000-volt PIV ratings are quite common and inexpensive. Projects 1, 2, and 3 use 50 PIV units, but you will probably find those rectifiers with the 1000-volt value to be no more expensive.

In Project 1, if you replace the transformer specified with one having a 300-volt secondary, you will end up with a no-load output voltage of approximately 500 volts dc. Rectifiers with a 1000-volt P.I.V. rating would then be required, and the working voltage rating of the filter capacitor would ideally be 600 volts or more. For this higher output one additional component *must* be added to the circuit for safety reasons. This is a bleeder resistor. The bleeder will also present a minimum load to the power supply, preventing extremely high swings of output voltage. Figure about 100 ohms-per-volt to be a good ballpark value for the bleeder resistor, so for 500 volts, a 50,000-ohm carbon resistor would be fine. A 2-watt power rating should prove adequate, although I like to increase the bleeder resistor power rating by as much as is practical. A 5-watt unit will assure that overheating and bleeder damage does not occur. When a bleeder resistor opens up, stored power remains in the filter capacitor, and at a 500-volt potential, it can be lethal given the right set of circumstances.

To accomplish the same (voltage and current) output characteristics with the full-wave center-tapped circuit of Project 2, the transformer secondary would have to have an output voltage of 600 volts across the entire secondary, or 300 volts from center tap to either side. Again, the rectifiers should be rated at 1000 PIV; the rest of the circuit can remain the same. The full-wave bridge circuit of Project 3 can use the same transformer as the half-wave circuit. Again, the diodes should be rated at the high value.

All of the changes discussed for the circuits of Projects 1, 2, and 3, with the secondary of the tranformer on out to the filter capacitor. Most likely, a few changes will be required in the primary portions as well. The degree of change will depend upon the power rating of the transformer and the amount of current which is to be drawn from the supply. At a nominal 12-volts output, a current drain of 1 ampere results in a power consumption from the ac line of only 12 watts. In ac line current, this is equivalent to approximately 100 milliamperes. The fuse listed is a half-amp type and will certainly not open under the maximum current drain which is equal to 1/5 of the total current it will pass. However, if we switch to an output of 350 volts dc at 1 ampere, as stated for the power supply modifications just discussed, this means a total of 350 watts of delivered power and a current drain from the ac line of about 3 amperes. Obviously, if the original half-ampere fuse remains, it will open. For the higher-rated power supply, use a 5- or 6-ampere line fuse. This will allow for a normal operating current of 3 amperes to pass but will open should a short-circuit occur.

Additionally, you might want to use a little heavier conductor in the primary circuit, since the drain from the ac line will be many times higher than before. Appendix B shows the amount of current which can be safely passed by various sizes of copper conductor.

Naturally, you can substitute any transformer into any of these three basic circuits to arrive at an output voltage which is suited to your individual needs. The no-load output voltage will be approximately 1½ times the secondary voltage (1½ times half the secondary voltage in a full-wave center-tapped configuration). Under heavy current drain, the dc output will be nearly identical to the RMS ac potential. Caution: The original circuit provides low-voltage outputs. As the secondary voltage of the transformer increases, so does the danger of severe electrical shock. Never omit the bleeder resistor from any power supply which offers anything but a low-voltage output. Additionally, be cautioned that any ac line-derived power supply is potentially deadly should anyone come in contact with a portion of the primary circuit. More persons are killed each year by 115-volt house current than by any other voltage source.

PROJECT 1
HALF-WAVE POWER SUPPLY

A half-wave power supply is not particularly efficient nor desirable at all for many critical solid-state circuits which require a stable dc source. However, the simplicity of this design and the ease of construction make it ideal as a power source for motors, fans, and simple dc devices which do not require an especially stable voltage output nor high amounts of current. Figure 6-1 shows the schematic representation of this simple circuit.

If you have studied the first five chapters of this book, you will immediately recognize the standard half-wave design, which incorporates a single silicon diode and an electrolytic capacitor at the transformer secondary. Notice that the primary circuit is safety-

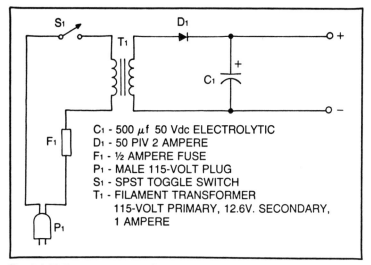

Fig. 6-1. Half-wave power supply.

equipped, in that a ½-ampere fuse (F1) is inserted in the ac line. Switch S1 is an SPST toggle type that is used for turning the supply on and off. S1 may be eliminated if desirable or may be replaced with the switching contacts from the switch in the equipment this supply will power.

This schematic specifies a 500-microfarad electrolytic capacitor with a 50 Vdc rating. This is a good value and will improve the regulation of the output voltage considerably over that which would be obtained with a filter capacitance value of 50 microfarads, for instance. If this supply is to be used for powering only electromechanical devices such as relays, fans, motors, etc., the value of C1 may be reduced considerably without any ill effects. Do not, however, decrease the voltage rating of this particular component.

Transformer T1 is a small filament type which contains a 115-volt primary and a 12.6-volt secondary. The 12.6-volt value is standard with some manufacturers, although some units will be rated at an even 12 volts ac. Either secondary voltage value is all right for this design.

Construction of this power supply is quite simple. You should be able to complete it in less than an hour, assuming that the cabinet or housing is already available. If you plan to mount the power supply in a metal box, drill a mounting hole for S1 and another for F1 if a through-the-panel fuseholder is used. Alternately, a fuse plug of an in-line fuseholder may be substituted for less mechanical complexity. Drill a hole in the metal compartment to allow for the exiting of the line cord which will ultimately connect to the 115-volt main. Fit this hole with a rubber grommet to insure that the insulation is not cut by the sharp edges. It's a good idea to use a three-prong line plug, connecting the ground contact to the metal chassis. This type is not shown in the schematic drawing and is not mandatory, especially when a plastic container is used as a cover or chassis instead of the metal one being discussed.

Connect the transformer primary leads to S1 and F1, as shown in the schematic drawing. Transformers are ac advices, so either primary lead may be connected to the two contact points shown. The transformer should be secured inside the power supply compartment. Many of these devices are fitted with pre-punched bases which will accept small nuts and bolts hardware for a firm mechanical connection.

It's now time to consider the secondary portion of this circuit. A terminal strip about three inches long may serve as the mounting platform for D1 and C1. Alternately, a small section of perforated circuit board may be incorporated, although more complex mounting considerations may be involved here. Figure 6-2 shows a method which uses the terminal strip. A standard five-pin strip is used, although only three of the pins are actually a part of the circuit. The central pin is often used as a ground point and has a small metal tab which is drilled to accept a small bolt. This can be used as the anchoring point to the chassis or compartment. Connect C1 as shown and then D1. The output conductor may also be connected at this time and then the far right contact and its elements soldered. The remaining connections can now be made and soldered as well. The two leads from the secondary of the filament transformer may be connected without regard to polarity. However, C1 and D1 are dc devices with specific polarities. A reversal of these latter two components can mean power supply failure or a reversal of polarity at the output of the supply. Make certain that the components in your circuit are connected as shown in the schematic drawing.

The power supply output leads may be brought completely out of the container or connected to feedthrough terminals. Alternately, these leads may be wired directly to the equipment the supply is to power. Before the latter is done, however, it will be necessary to check the operation and performance of the finished circuit.

Before connecting P1 to a 115-volt source, closely re-examine the entire project from start to finish. Make certain all contacts have been properly soldered; that there are no short circuits, either

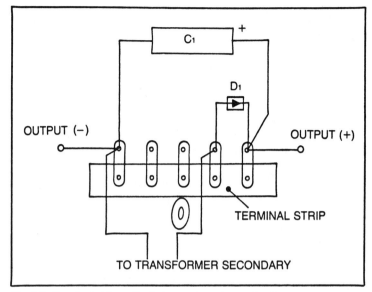

Fig. 6-2. Terminal strip method of power supply construction.

between components or between the circuit and a metal chassis. Re-inspect the diode and capacitor, checking for polarity reversals. Make certain you have not reversed the primary and secondary leads of the transformer. Should you accidentally connect the transformer secondary to the 115-volt line, F1 will open up, but probably not before D1 is destroyed by an instantaneous high voltage that will result. Pay special attention to the primary circuit around F1 and S1. If any shorts occur between these devices and P1, a possible fire hazard could still exist if the house fuses in the portion of the home electrical system powering this circuit are overrated and fail to open up. This is a very rare occurrence and is mentioned here only to stress the fact that proper building techniques and safety considerations should always be in the forefront.

If everything seems to be satisfactory so far, make certain S1 is open and insert the line plug into the wall outlet. At this point, absolutely nothing should occur. If you hear a humming or see any signs of smoke or electrical arcing, immediately disengage the line plug and check your primary circuit. If nothing happens when P1 is inserted, connect the probes of a voltmeter across the output terminals of the supply. Make sure the positive probe is connected to the positive lead. The same applies to the negative probe and the negative power supply lead. Set the voltmeter to a scale which will read at least 20 volts dc. Now, while observing the meter, activate S1. The meter should immediately climb to a value of between 15 and 20 volts dc, the peak value of the 12.6-volt transformer secondary output. Once the needle has reached its peak, it should remain steady with no further up and down movement. Switch off the power supply by disengaging S1 and continue to watch the meter. This power supply does not contain a bleeder resistor simply because the electromechanical load will quickly discharge C1. However, in this test situation, the only load the power supply is connected to is the voltmeter. This instrument is designed to present an extremely light load to any circuit it measures, so if your supply is working properly, disengaging S1 will create no immediate change in the indicated voltage. Your power supply is completely removed from the ac power source, but the charge in the capacitor is still supplying power (voltage and current) to the meter.

Even though the voltmeter represents an extremely light load, current is still being drawn by its internal resistors, and within a short period of time, you will see the voltage indicator begin to drop. In no time at all, all of the power from C1 will have been discharged, and the supply will be truly inactive again. This test not only assures the proper operation of your power supply but also demonstrates very effectively the power storing capabilities of capacitors in power supply circuits, which were mentioned in an earlier chapter. The dangerous aspects of this can be appreciated if you imagine that the stored potential is more on the order of 1800 volts than 18, as is the case in this project.

After you have obtained a proper reading on the voltmeter and have deactivated S1, should the voltage drop immediately, this would indicate a problem in C1. It might be reverse-connected in the circuit, not connected at all or defective. The meter would then be indicating the pulsating dc output from the rectifier alone. This, however, would cause the indicating needle to flicker and should be apparent before S1 is deactivated.

If, on the other hand, when S1 is activated, you get no reading whatsoever, immediately switch S1 off again and remove the plug from the wall. Check to make certain that the voltmeter probes were correctly placed at the power supply output terminals. If they were reversed and the power supply was working properly, the voltmeter indicator would dip below zero.

If you are certain that the voltmeter is properly connected, then remove F1 to see if it is open due to a high current drain. Caution: Do this only with the line plug removed from the wall. Voltage will always be present on F1 when the plug is connected to a power source, regardless of the position of S1. If the fuse has blown, this is an indication of a short-circuit either in the primary or secondary portion of the supply. A defective diode or capacitor can sometimes cause this to happen, or even an internally shorted transformer winding. The defect will certainly be found by checking all wiring and the quality of each of these components.

If the fuse is not blown, this is an indication of an open circuit. This can be caused by faulty or broken solder contact, a broken conductor, defective transformer, capacitor or diode. Re-examine all wiring and re-check all components, and the problem will most likely be found.

If no apparent defect makes itself known, re-insert the plug into the wall outlet and activate S1. With an ac voltmeter, check to see that the transformer primary is actually receiving the proper current. This is indicated by a 115 Vac reading at the primary input leads. Next, check the transformer secondary output ahead of the diode. Here, you are looking for a reading of about 12 Vac. If you have nothing here, this indicates a defective transformer, assuming that voltage was measured on the primary windings.

If the secondary checks out all right, switch your meter to dc voltage again and check the output from the diode by temporarily removing the capacitor from the circuit. If you get no reading at this point, the diode is defective and must be replaced. If you do get a reading, reconnect C1. Take measurements again from the same point. If no voltage is indicated, C1 is defective or shorted. If you do get a reading, then the output conductor between C1 and the positive terminal is open.

This is an extremely simple power supply, and these detailed troubleshooting instructions will probably not be required at all. When trouble does develop in a power supply of such non-complex design, the problem area is usually quite evident. However, this has been a good project to serve as an example for detailed troubleshooting procedures, because the technique is so much more easily explained. The instructions given here (with some modifications) can be used to troubleshoot nearly every type of basic power supply and may come in handy in dealing with other projects in this chapter.

Once your power supply is working properly, the finishing touches may be added to make it ideally suited to individual needs. You may wish to install a 115-volt panel lamp somewhere in the circuit to serve as an indicator when power has been applied by activating S1. Some lighted switches are even available from most hobby stores which will glow whenever power is applied.

The output from this supply will vary from about 12 volts at maximum current drain (about 750 milliamperes) to nearly 20 volts under light loading. If you find the voltage to be too high, try inserting a resistor in series with the positive lead. A variable resistor is even better, as it will allow you to adjust for the proper voltage while a load is connected. The value of the resistor will depend upon the voltage drop desired and the current drawn. If the device to be powered draws 100 milliamperes, then you will get an approximate 1 volt drop for every 10 ohms of resistance.

This completed half-wave supply costs little, is certainly easy to construct and test, and can serve a myriad of non-critical power supply requirements for electromechanical and simple electronic devices. Do not attempt, however, to use this supply with critical electronic circuits such as ICs, microprocessors, etc., as the instability of the nonregulated voltage will create operational problems. For these latter devices, you will find more suitable power supplies later on in this chapter.

PROJECT 2
FULL-WAVE CENTER-TAPPED SUPPLY

The previous power supply project provided an output of a little over 12 volts under a moderately heavy load and it was quite simple to build. However, all half-wave circuits have built-in limitations which make them suitable only for the most non-critical applications. The power supply circuit which has been most often used is the full-wave center-tapped design. This rectifier configuration allows operation and rectification from both halves of the ac sine wave. This increases the ripple frequency and makes filtering far less difficult. The full-wave center-tapped configuration is not appreciably more complex than its half-wave counterpart. All that is required is a transformer with a center-tapped secondary and an extra diode. The total ac value of the entire transformer secondary must be equal to roughly twice that desired as the dc voltage output. Full-wave center-tapped power supplies have identical primary circuits to those of half-wave design and may still be filtered by a single capacitor.

You can readily see all of this in Fig. 6-3, which is a full-wave center-tapped power supply that delivers an output voltage equivalent to that of our former half-wave supply. Indeed, most of the components for this project are identical to those of the last one.

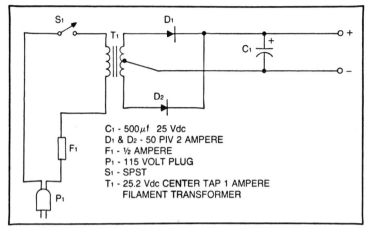

Fig. 6-3. Full-wave center-tapped power supply.

Transformer T1 has been changed to a 25.2 volt dc center-tapped type and D2 has been added. Other than this, the circuit is basically the same as before.

Construction may take place in exactly the same manner as for our half-wave supply, although it will be necessary to alter the terminal strip wiring slightly as shown in Fig. 6-4. Here, we can see that C1 and D1 are connected as before. The output terminals remain the same (1 and 5). The transformer center tap now connects to terminal 1, while the other two transformer leads are attached to 2 and 4, respectively. Silicon diode D2 is added between terminals 2 and 5. These are the only changes required. The same 500 microfarad filter capacitor is used, although this value may be halved and the supply will still have as good a voltage regulation factor as was obtained with the half-wave supply using higher capacitance values.

Instead of using the terminal strip design, you may wish to lay out the components on a small piece of printed circuit board. Circuit board construction is to be preferred for rectifier and filter circuits as the components become more numerous. Figure 6-5 shows a typical layout which may be had with a 4" × 4" section of circuit board that can be purchased from any hobby store. There is adequate room left over to add a zener diode regulator circuit at a later time, if so desired. Such a circuit is presented later in this chapter.

Since there is another diode leg in a full-wave center-tapped circuit (to act on the remaining half cycle), wiring is a little more

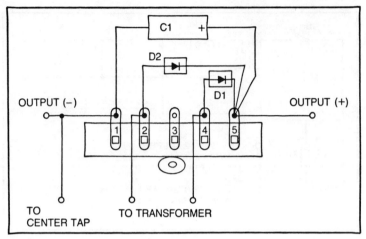

Fig. 6-4. Terminal strip wiring of full-wave supply.

complex, but not appreciably so. Once you have completed the project, examine the wiring very carefully to make certain the polarities of D1, D2, and C1 are correct. Also, be certain the two outside leads of the power transformer are attached to D1 and D2 and that the center tap goes directly to the negative contact of C1.

If all appears to be in order at this point, make sure S1 is in the off position and insert P1 into the wall outlet. As before, nothing should happen until S1 is activated. When this is done, the attached voltmeter should read about 18 Vdc. Again, there should be no wavering and the voltage should decay shortly after S1 is disengaged. It should not drop off immediately.

Should the power supply not function, perform all of the checks outlined in the troubleshooting discussion offered previously. There are a few modifications which must be made when troubleshooting the secondary portion of this circuit. If the primary seems to be operating properly, attach an ac voltmeter across the outside leads of T1. A reading of approximately 25 volts should be obtained. Leaving one probe in place, move the other to the center tap. Here, a reading of approximately 12 volts should be noted. Leaving the center tap probe in place, lift the remaining probe and place it to the opposite transformer lead. An identical reading to the previous one should be obtained at this point. From this point on, troubleshooting procedures are nearly identical. Lift C1 from the circuit temporarily, and with the voltmeter in the dc mode, measure the pulsating dc voltage between the D1-D2 junction and the transformer center tap. If you get nothing here, then either or both diodes may be defective. If readings are normal, connect C1 again. If the reading is normal, then the conductor between C1 and the output terminal is open. If the reading goes to zero, then C1 is defective or shorted.

The troubleshooting procedure for this full-wave center tapped supply is very similar to that for discovering problems in a half-wave circuit. The full-wave center-tapped supply is simply two half-wave supplies combined in parallel.

The output from this power supply may be used to drive all of the devices which were applicable to the half-wave supply. Additionally, some solid-state electronic devices such as transistor radios, some hand-held computer toys, and other non-critical circuits will find the output voltage perfectly accessible. As before, it may be necessary to install a series resistor in the positive output lead to drop the voltage. This will depend upon the current drain and the voltage drop which is needed.

This is a good basic power supply and may be fitted with outboard regulators, meters and a number of other accessory devices to make it very valuable in many critical applications. Some of the power supply projects to follow will use this basic design with built-in regulators to provide computer-grade stability. The author recommends that this supply be mounted in a small aluminum box. The center tap of the transformer may be connected directly to the aluminum. Here, the chassis itself serves as circuit ground. The negative lead from C1 may also be attached to the chassis. Now, the metal case will be an actual voltage point (negative). Only the positive lead will require a feedthrough connector. For increased

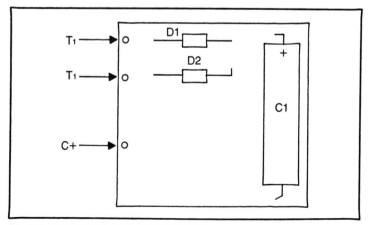

Fig. 6-5. Component layout on circuit board.

stability, a 5 kohm ½-watt carbon resistor may be wired in parallel with C1, although this is not necessary for most non-critical applications.

PROJECT 3
FULL-WAVE BRIDGE SUPPLY

Staying within the same area of study as with the previous two projects, the supply shown in schematic form in Fig. 6-6 provides the same output voltage, current and regulation characteristics as the full-wave center-tapped supply. The only difference in the two is that the full-wave bridge circuit uses a different rectifier configuration and a transformer which is identical to the one used in the half-wave supply (115-volt primary; 12.6-volt secondary).

As you learned in a previous discussion, a full-wave bridge supply acts upon both halves of the ac cycle and offers the same filtering ease (due to a high ripple frequency) as that of the full-wave center-tapped design. The full-wave bridge circuit has become more and more popular in recent years. This is attributed directly to solid-state rectifiers. Several decades ago when vacuum tube rectifiers were widely used, the bridge circuit was not as attractive because twice as many rectifiers are required as with the center-tapped circuit. Vacuum tubes require separate filament supplies, give off heat, are quite expensive (when compared with solid-state rectifiers), and present rather complex mounting problems.

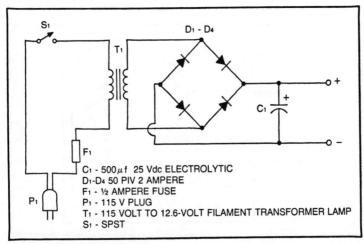

Fig. 6-6. Full-wave bridge supply.

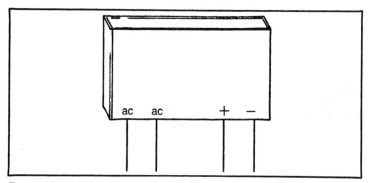

Fig. 6-7. Solid-state bridge rectifier package.

A full-wave bridge power supply might be preferred today because of economy alone. Silicon rectifiers can often be purchased for ten cents each from surplus outlets, so cost in this area is not a big factor. On the other hand, the power transformer is usually the most expensive item in any basic power supply (using solid-state rectification). Using the projects already discussed, a transformer with a 12-volt secondary winding is usually a bit cheaper than one with a 25 volt secondary. The transformer may also be a bit smaller. Therefore, if we can obtain the same output voltage and regulation characteristics of a full-wave center-tapped circuit by using a solid-state bridge and a less expensive power transformer, then it is probably best to go this route for most applications.

Look again at Fig. 6-6. Notice that the primary circuit is identical to the previous two, as is the output circuit past the bridge rectifiers. Also notice that the transformer contains no center tap connection. The two secondary leads connect to the center portion of the bridge, whose output is fed directly to our old faithful, 500 microfarad, 25-volt electrolytic capacitor.

Here, it is best to use a small section of circuit board as the mounting platform for the four silicon diodes. The only complex point in construction is in wiring these four diodes to match the configuration shown in the schematic drawing. Also, it is mandatory that the transformer leads be connected as shown and that the negative leads be connected to the back of the bridge and the positive from the front. The filter capacitor (C1) is connected across the output of the power supply as before and smoothes out the pulsating direct current.

There is one modification that you can make to this circuit which will save you a bit of space and will also reduce the time required for completion. Figure 6-7 shows a drawing of a solid-state bridge rectifier integrated circuit which is inexpensive and available from most hobby stores. What is contained inside of this package is four separate silicon diodes wired in the bridge configuration. The correct leads are brought to the outside of the package and marked for connection to the power supply and to the output circuitry. The example shown is one of many. Typically, the ac leads are identified with the letters "AC". Alternately, they may be marked with the sine wave symbol ($\sim$). The positive and negative output leads are almost always identified with plus and minus symbols, although occasionally, a red circle will be used for identification of the positive lead alone. The unmarked lead is then assumed to be negative.

If you buy your silicon rectifiers from a surplus house, you may find that it is cheaper to go this route. If, however, you will be buying them from a hobby store, you will probably find that it is cheaper to purchase a single package bridge circuit than it is to pick up four discrete silicon rectifiers. One other consideration is in mounting. Using the solid-state package, you can probably do away with the perforated circuit board and use the terminal strip or even direct wiring to the power transformer. Here, it may be necessary to fasten the rectifier package to the side of the transformer case with a drop or two of epoxy cement. Make certain all of your connections are correct, however, before performing this permanent securing operation.

There is no electronic advantage to using the single rectifier package over four discrete diodes. Some would argue that there is a bit of a disadvantage, in that if one of the internal diodes in the package fails, it will be necessary to purchase a new unit. Using discrete components, the failure of a single diode means replacing only one component, leaving the good ones in place. Actually, the cost is so low that it is really pointless to argue the matter. I usually opt for the solid-state package to facilitate construction ease and reliability.

After your project is completed, make the mandatory visual inspection which is required of all electronic construction. Sometimes, it's a good idea to grasp a conductor and shake it gently, observing the solder contact point for any signs of movement. Your inspection should be especially attuned to discovering any suspicious looking solder joints. A rough, dull surface or a flake-like appearance is often a sure sign of a bad connection, one which will lead to future, if not immediate, problems. With S1 in the off position, insert the line plug and go through the same procedures as before. If you fail to get a reading on the voltmeter, be highly suspicious of the rectifier assembly, especially if discrete diodes were used. If you have reversed a diode, the circuit will not work and you may have even damaged one or more of these solid-state components. A resistance check in both directions with an ohmmeter will quickly identify the bad unit. (Make sure this is done with the line plug removed from the wall outlet.) Naturally, you will want to conduct the standard troubleshooting procedures, starting at F1 in the primary circuit and continuing on through the secondary.

The output voltage from this supply should be identical to that of the full-wave center-tapped circuit. It offers good dynamic regulation (for a basic supply) with a 500-microfarad capacitor. If you require better regulation, try paralleling another capacitor of the same value with C1. Be certain to observe polarity, because a mismatch at this point will result in component damage.

The full-wave bridge power supply may be used to power many different types of electromechanical and electronic circuits. With the inclusion of a series resistor, transistor radios will perform in exactly the same manner as they do with batteries. The circuits which will not do as well will be those whose current demands vary over a broad range. Here, the voltage may tend to fluctuate, especially when these demands change from very low to very high on a repetitive basis. When this occurs, the voltage changes often induce unwanted effects within the more critical pieces of electronic equipment. Most radio frequency circuits cannot be operated properly from this supply if the voltage is used in the frequency-determining portions of the device. For these applications and others, an outboard regulator circuit would be necessary.

PROJECT 4
ADD-ON ZENER DIODE REGULATOR

We can use any of the three previous power supplies to provide operating potential to electronic devices which are more demanding in voltage stability by adding an accessory to the original circuits. One of the simplest forms of electronic regulation involves a passive device known as a zener diode. The zener diode has been discussed previously. In a regulator circuit, the zener diode acts with a series resistor to maintain voltage at a fairly precise level. When the power supply output voltage is at or below the zener value or zener "knee", this component does absolutely nothing. But when the voltage rises above this certain value, the diode conducts. This creates a temporary short-circuit at the power supply output and more current is drawn. (Actually, the short-circuit is only a partial one.) As the current drain increases, voltage is dropped across the series resistor and the total output voltage is brought back to the zener knee level. This process occurs many thousands of times within a normal operating cycle, and the result is a stable output voltage.

Figure 6-8 shows the simple zener diode regulator circuit, which consists of only two additional components, the series resistor R1 and the zener diode CR1. Notice that the zener diode is shunted across the original power supply output (between the positive and negative leads). It can be seen that when the zener conducts, a partial short-circuit occurs, causing more current to be drawn through R1 which will, in turn, drop the output voltage.

This particular circuit specifies a 12-volt half-watt zener diode. This means that the output voltage will be held at a constant 12 volts dc. If power drains are not to be excessive, you could also use a 15-volt zener diode in order to arrive at a slightly higher regulated output voltage. While the regulator circuit may be used with the half-wave power supply circuit, this is not especially efficient. If you have a need of regulated output voltage, then the full-wave circuit (either center-tapped or bridge) should be incorporated for ease of filtering and better overall regulation.

The resistor specified here is a 180-ohm 1-watt carbon type. If you desire a 15-volt output, it may be necessary to decrease this value to 120 ohms. You can even get a 5, 6, or 9 volt regulated output by opting for zener diodes of these values and increasing the ohmic value of R1. For these latter values, it will be necessary to choose zener diodes of the 1-watt variety (instead of the half-watt specified), because of the additional current which must be passed when operated in the low-voltage mode.

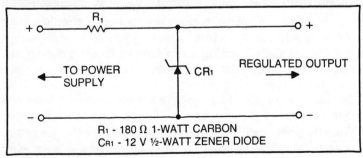

Fig. 6-8. Zener diode regulator circuit.

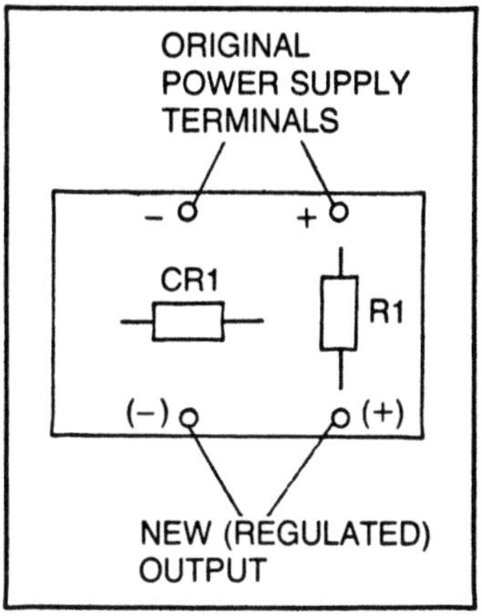

Fig. 6-9. Mounting of regulator in outboard fashion.

```
ORIGINAL
POWER SUPPLY
TERMINALS
       -o   +o
         CR1
       ──[▭]──   [R1]
       (-)o        o(+)
         NEW (REGULATED)
         OUTPUT
```

The zener diode is a polarized device. This circuit will not operate if this component is reversed. Pay careful attention to polarity when wiring this circuit. You may decide to include this circuit with one of the basic power supplies already discussed, building it into the original circuitry. Alternately, you may decide to use this regulator in "outboard" fashion. Figure 6-9 shows a convenient way for this to be accomplished.

A small section of perforated circuit board is drilled at the top center to accept the original power supply output terminals. This assumes the screw-in type of terminals have been used. The circuit board is simply mounted to the terminals and held fast by the screw-down nuts. At the bottom of the circuit board, two more holes are drilled through which two additional screw-down terminals are mounted. These are connected to the positive and negative outputs of the regulator circuit. The series resistor and zener diode are wired between the two sets of terminals. Notice that there is no permanent or soldered connection between the original power supply and the new regulator circuit. Now, the two new terminals will provide a regulated voltage output at a total additional cost of less than three dollars (if you shop wisely). If you want to return to the non-regulated version again, all you need do is quickly unscrew the top two terminal nuts and remove the regulator circuit board. Alternately, you may simply connect to the top two terminals for an unregulated output, although it is best to remove the regulator circuit entirely, since it will still draw some current and affect even the unregulated output.

You may wonder why we have not designed a separate case for this regulator add-on. This is because we are dealing only with a low-voltage potential. No primary circuit wiring is involved here, and the maximum potential which is exposed is a safe 18 Vdc (at the unregulated output terminals of the supply). Certainly, you may wish to make arrangements to install this circuit board within your present power supply and provide two sets of output terminals (regulated and unregulated) for connection to devices to be powered. Here, it is a good idea to install a switch between R1 and its power supply input in order to disable the portion of the circuit used for regulation. This will prevent wasteful current drain during non-regulated operation. Figure 6-10 shows the placement points for this deactivation switch.

Checkout procedure is very simple, assuming that it has already been established that the unregulated power supply is fully operational. With the power supply deactivated and the regulator circuit board in place, connect the probes of an accurate dc voltmeter across the regulator output terminals. Make certain you observe polarity and that the probes are not reversed. With the voltmeter set to a range which will measure 12 (or 15) Vdc, activate the power supply and note the reading. It should be of exactly the same value as the zener diode. If you get no reading, quickly take a measurement of the unregulated output voltage. If nothing is indicated, you have a problem with the original power supply. If the reading is normal, deactivate the power supply and recheck the regulator circuit. There is a very good chance that you have reversed the zener diode, have a defective series resistor, or a broken conductor somewhere in the assembly.

If the output from the zener regulator is far above the design value, then the zener is probably defective and should be tested and replaced. Chances are, however, if you use functioning components, your circuit will work correctly the first time and you will now be the proud owner of a power supply which offers an electronically regulated output.

The zener diode-regulated power supply may be used to provide operating power to small transmitters, radio receivers, computer games and a myriad of other electronic devices. Zener diode regulation is the simplest form of electronic voltage control and while providing a highly stable output as compared with non-regulated supplies, may still not be adequate for highly critical, frequency-sensitive circuits. These latter devices will require sophisticated electronic regulation using active solid-state elements. These will be discussed later in this chapter.

While your zener-regulated supply will not do everything, it can be used for powering most integrated circuit devices, solid-

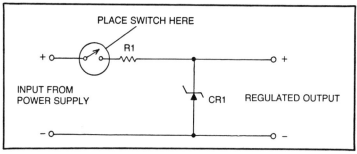

Fig. 6-10. Placement points for deactivation switch.

state electronic circuits and electromechanical units with great efficiency. The fact that the voltage is held to a nominal value is a great help in establishing proper operating conditions and in figuring total power output when this is necessary. The output current rating from the regulator will be approximately 40 milliamperes.

PROJECT 5
DUAL VOLTAGE POWER SUPPLY

Many types of electromechanical and electronic equipment will require more than one operating voltage potential. Here, you can build two separate dc power supplies. Of course, this involves about twice the expense of a single circuit, increases mounting space and possibly circuit reliability. Alternately, you can build an economy power supply which uses only a single transformer but produces two voltage outputs, one at twice the other.

The economy power supply circuit is shown in Fig. 6-11 and appears at first to be a standard full-wave bridge rectifier. A closer inspection reveals the center tap connection, but the center tap is not grounded as before and serves as the positive low-voltage output point. Is this circuit of full-wave bridge design or full-wave center tap? The answer is both. What we have done is to connect a full-wave bridge circuit to the outside leads of the center-tapped transformer for a standard dc output. But we have also used the center tap connection to obtain an output voltage which is half that of the bridge value. The center tap is used as the positive connection, because the back two diodes of the bridge circuit are used for rectification of the low voltage output. The center tap is positive in relationship to circuit ground. Notice that a separate filter capacitor is used for each output leg. The posted output voltages can be considered as absolute minimums using the components specified.

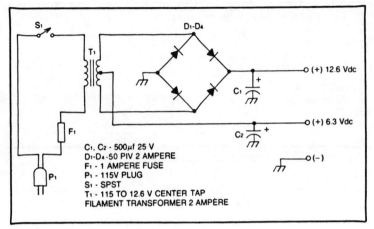

Fig. 6-11. Economy power supply circuit.

Under heavy load currents, the voltages may drop to these points, but will tend to be about 50% higher than posted under light or no-load conditions.

The primary circuit is similar to the previous power supplies, although a 1-ampere fuse has been used here, owing to the increased current rating of the transformer. Since both power outputs derive from the transformer, their total current drain adds. If 500 milliamperes of current is drawn from both sources, then a total current drain of 1 ampere will be seen at the secondary. A 2-ampere rating was chosen so that the current drain from each of the output terminals would be close to that of the previous power supply projects.

The construction of this power supply is no more difficult than any of the previous projects. You can start by simply building a standard full-wave bridge power supply, although you will be using a center-tapped transformer. Just make certain that the two outside transformer secondary leads connect to the proper points in the bridge rectifier circuit and all will be okay. Certainly, you may use an integrated circuit type of bridge assembly or elect to use discrete components instead. This was discussed earlier. Either will work equally well.

After you have completed this portion of the supply, run through the test procedure described for Project #3, and make any repairs or adjustments necessary to obtain proper operation. Once the bridge portion of the supply is working, simply connect another filter capacitor (C2) across the center-tapped transformer lead and reactivate the supply for another measurement. You should get a reading equal to exactly half of the higher voltage value between the transformer center tap connection and circuit ground. If the standard bridge circuit is operational, then the low voltage circuit must be as well, unless you have a shorted filter capacitor or broken center tap connection.

The economy power supply circuit does not offer two discrete power supplies. It does offer two different voltage outputs; but remember, they are derived from the same source. Using this supply, a high current drain from one output can result in a voltage drop at both. Since the transformer secondary is designed for maximum ac output current drain of 2 amperes, this will mean that the maximum total current drain from the dc outputs should be no more than a total of 1500 milliamperes (1.5 amperes). So, you may draw 750 milliamperes from each dc voltage output or 1000 milliamperes from one and 500 from the other. As long as the total dc current drain stays at 1500 milliamperes or less, operating maximums are being observed.

You can make this power supply more versatile by installing series resistors in each of the output lines. This is shown in Fig. 6-12. The value of R1 and R2 will depend upon the amount of current which is to be drawn from each leg of the supply and the amount of voltage drop desired. If each leg is drawing 500 milliamperes, then a series resistance value of 2 ohms will produce a 1-volt drop. The unloaded output of the higher voltage line will be about 18 Vdc. Under light loading, this will drop no more than 1 volt. If you desire a nominal 12 Vdc output to a device which draws 500 milliamperes, you will probably have to have a series resistor value of about 6 ohms. Under these current demand conditions, a variable resistor with a 0 to 10 or 15 ohm range might be ideal. Alternately, a 0 to 50 ohm unit would offer more versatility in different uses, as lower current drains will require a higher series resistance. If you know the operating current of the device to be powered by this supply,

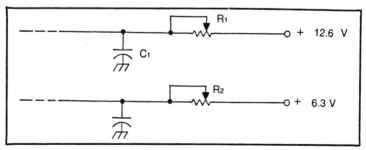

Fig. 6-12. Installation of series resistors in output lines.

your job is much simpler and a fixed resistor which is far less expensive than the variable type may be inserted after certain calculations are performed. If you require a drop of 5 volts at an operating current of 100 milliamperes, you can use Ohm's Law (E = IR) to quickly find out the resistance value. Here, E is the voltage drop, while I is the operating current and R is the resistance. E is given in volts, I in amperes and R in ohms. The basic formula as shown is really designed for figuring voltage when current and resistance is known. The modification of this formula to solve for unknown resistance is $R = \dfrac{E}{I}$, or resistance equals the voltage drop divided by operating current. So, by inserting our known values, we arrive at:

$$R = \frac{5}{.1} \text{ or } R = 50 \text{ ohms}$$

Looking at the formula, you can see that I is stated as .1 because this value must be given in amperes. A 100 milliampere current drain is equivalent to 0.1 amperes.

We now know that in order to arrive at the desired output (which will be brought about, in this case, by a 5 Vdc voltage drop), we must insert a series resistor with a value of 50 ohms. The same formula may be used for any other voltage drop to any power supply circuit which is not electronically regulated.

We must also figure the power rating to be applied to the series resistor. Again, we fall back on the Ohm's Law power formula, which applies to this particular situation. It reads: $P = I^2R$, where P is power dissipation in watts, and I and R are, again, current and resistance in amperes and ohms, respectively. Our formula for figuring the power which the series resistor must dissipate would then read:

$$P = .1^2 \times 50 \quad \text{or} \quad P = 0.5 \text{ watt}$$

The formula tells us that the series resistor must be capable of dissipating ½ watt of power. Half-watt resistors are quite common, especially in 50-ohm values, but so are 1-watt units. While it would appear that a half-watt power dissipation factor in the series resistor would be adequate, the unit would be operating at maximum ratings. It would be safer to use a 1-watt unit at a 50% derating factor to improve reliability.

This example has demonstrated the proper methods of figuring series resistances for insertion in the output leads of power supplies in order to effect a useful voltage drop to match the operating needs of electronic loads. Remember, however, that a series resistor will present a specific voltage drop only at one output current value. If the current drain from the supply drops, the voltage will increase. An increase in output current will cause a further drop in voltage and increased power dissipation. Therefore, a series resistor is mostly used in power supply output leads when the total load remains constant.

If your economy power supply is to be used to power electronic loads which require stable operating voltages even though the load demand will vary continuously, you might want to consider adding an outboard regulator circuit (or circuits), as was described in the last project. For the higher voltage supply, Project #4 can be used directly as is. By changing the zener diode to a 6.3 volt unit (or a similar value), the low voltage power supply leg can also be regulated. Some applications may make it desirable to regulate one output while leaving the other in its unregulated state.

As has been the case before, the power transformer specified in this project may be changed to deliver higher or lower output voltages. The same caution which was discussed earlier applies to this and all other supplies when output voltages surpass small values.

The schematic shown for the economy power supply circuit is a general one. I used the one shown in Fig. 6-13 to supply power to a piece of war surplus equipment. Using the values shown, this supply delivered an output of 15 Vdc and 7.5 Vdc under moderate current drain of 350 milliamperes (approximately) from each leg. It was not necessary to insert any series resistors and electronic regulation was not required. Two bleeder resistors, however, were added to present minimum loads on the supply and to prevent swings to maximum peak voltage.

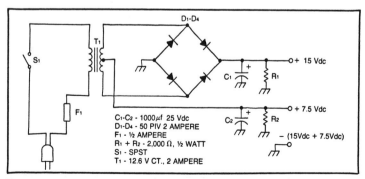

Fig. 6-13. Economy power supply schematic drawing.

157

PROJECT 6
DUAL-POLARITY REGULATED 15-VOLT SUPPLY

Many types of linear integrated circuits require a dual-voltage input in order to operate properly. This means the power supply must deliver a positive voltage in respect to circuit ground, and simultaneously, another voltage which is negative with respect to circuit ground. Additionally, good regulation is often required to prevent undesirable effects within the IC. A negative power supply is identical to a positive power supply. The only difference is in the former, the positive terminal serves as ground instead of the other way around. A dual-polarity supply might consist of two separate circuits, each with its own discrete components (transformer, rectifiers, etc.), or it may be an economy type which is slightly different from the project previously discussed.

Figure 6-14 shows the schematic diagram of a dual-polarity power supply designed to produce two separate voltage outputs with respect to circuit ground. One output is positive 15 Vdc; the other is the same value but at reverse polarity. Remember, these plus and minus values are with respect to a common ground. Notice that a center-tapped power transformer is used, along with what appears to be a standard bridge rectifier circuit. This may be a bit confusing, but this standard bridge is not used in its normal configuration but as two full-wave center-tapped rectifier circuits. This can best be explained by observing the two right-hand rectifiers in the stack (rectifiers 3 and 4). If you just leave these and remove

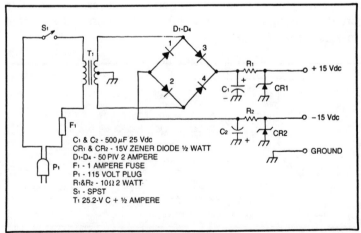

Fig. 6-14. Dual-polarity power supply.

rectifiers 1 and 2 along with C2 and the remaining circuitry out to the -15 Vdc output, you have a standard full-wave center-tapped circuit identical to the one discussed as Project #2.

By the same token, if you remove rectifiers 3 and 4 and all of their associated circuitry, you have another full-wave center-tapped supply left over composed of rectifiers 1 and 2, C2, R2 and CR2. Notice, however, that the rectifiers are reversed from previous examples. This is because they are to deliver a negative output voltage with respect to the center tap, which is circuit ground. Notice also that C2 and CR2 (polarized devices) are reversed in the lower positive leg when compared to the top positive one.

This supply looks very similar to our dual-voltage project discussed previously, but notice that the back of the rectifier stack is not grounded, while the center tap is. If you were to connect the top of C2 to the center tap (after it has been ungrounded) and ground the connection point of rectifiers 1 and 2, you would arrive at the previous dual-voltage supply. Dual voltage is not what we are concerned with here so much as dual polarity, although we do indeed have two separate voltage lines. Both outputs are at the same potential, but one is negative and the other positive.

Most linear integrated circuits require a reasonably well-regulated power supply. This is provided in this circuit by installing two zener diode regulators which are nearly identical to the previous outboard project discussed. To avoid confusion, let's examine the positive circuit leg only. The rectified output from diodes 3 and 4 is fed to filter capacitor C1, where it is smoothed. The filter output then enters the electronic regulator circuit comprised by series resistor R1 and zener diode CR1. The zener diode is a 15-volt, ½ watt unit. The resultant regulated output is held at a constant positive 15 Vdc.

The exact process occurs, but in reverse polarity, in the remaining leg of the circuit. The rectified negative output from the transformer secondary comes off the junction of diodes 1 and 2. Here, filter capacitor C2 takes over. Notice that its negative terminal is connected to the output line, while its positive terminal is grounded. The filter output is then connected to the negative regulator (notice the reverse connection of CR2), and the final output is a stabilized negative 15 Vdc. The transformer center tap connection serves as circuit ground for both output lines.

The primary circuit wiring is pretty much standard and identical to previous projects. The secondary is not too unusual, in that the full-wave bridge rectifier configuration (although not used as such) is laid in a normal manner. Since this is a dual-polarity power supply, it is especially important to pay strict attention to the polarity of each individual component. This includes the four solid-state rectifiers or an IC equivalent, the two filter capacitors, C1 and C2, and the zener diodes, CR1 and CR2. Remember, C1 is connected to the output line in reverse configuration to C2. This same reversal applies to CR1 and CR2. Improper polarity of any of these components will result in improper operation or complete circuit malfunction.

The entire secondary circuitry should be constructed on a small piece of perforated circuit board, which may be purchased from Radio Shack or another electronic hobby outlet. A 5" × 5" section should be more than adequate, with the physical height of the filter capacitors being the main determining factor in final dimensions. Some capacitors may be slightly larger than others. If space is at a premium, try to locate some miniature versions. If not, you can use units which are rated at more than 25 Vdc. Figure 6-15 shows a good circuit board configuration to use. This physical layout closely follows the logical schematic order of progression, in that the diodes are mounted at the transformer input. The capacitors are located to either side. Notice their reversal regarding polarity and position on the circuit board. Between the two are the half-watt series resistors, followed by the two zener diodes. The zener diode output leads are connected directly to the positive and negative output terminals. The ground terminal also serves as the center tap connection. It should be possible to wire this circuit board in a fairly short period of time, although you are cautioned not to rush because errors may be increased.

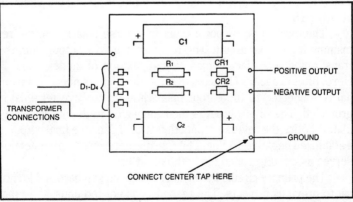

Fig. 6-15. Circuit board component configuration for dual-polarity power supply.

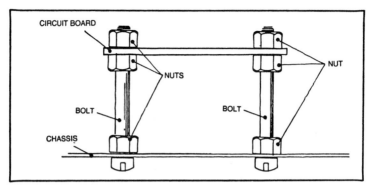

Fig. 6-16. Preparation of power supply chassis.

After all of the components have been added to the circuit board, visually inspect your component placement, connections, polarity and especially all solder joints. The circuit board wiring is a little more complicated than was the case with previous projects, but this still can be considered to be a very simple layout. If all appears to be in order, the circuit board may now be mounted on a chassis or in a case which also houses the power transformer. Mounting of the circuit board is very simple. All that is necessary is to drill two holes in the circuit board, one at the center of each end. Alternately, four holes may be drilled, one in each corner. The holes should be made away from the conductors or electronic components in an unused portion of the board. Hole diameter should be adequate to accept a firm fit from a small 1½ inch bolt.

Referring to Fig. 6-16, drill matching holes through the chassis or compartment which is to house this circuitry and insert the bolts as shown. A total of three nuts are used on each bolt. The bottom one secures the bolt to the chassis, while the remaining two are sandwiched around the circuit board. The bolts then form support legs to keep the circuit board firmly isolated above the case. There are other methods which may be used for mounting, but this one seems to be the most economical. It's simple as well as inexpensive and uses readily available hardware.

Once the circuit board is mounted firmly, connect the power transformer to the appropriate terminals and connect the output of the circuit to matching terminals as well. Your circuit should be complete in every detail. Before locking down the compartment cover, however, we must test the circuit to make sure that operation meets design standards and that no further corrections or modifications are necessary.

Begin with the positive voltage output leg. The positive probe of a voltmeter designed to read 15 Vdc or more is connected to the positive terminal, while the negative probe is placed across circuit ground (the center tap connection). Activate S1 and observe the meter for the voltage reading. The indicator should be exactly on 15 Vdc, although the reading may be slightly higher or lower than this due to voltmeter inaccuracy. If you get no reading, then it will be necessary to check the line fuse in the primary circuit and also to make sure that the line plug is connected and primary current is being delivered. If all appears well in the primary circuit, then remove the probes and place the negative one at the negative output terminal and the positive one to ground. If you still get no reading, then the power transformer may be defective, two or more diodes are damaged or there is a broken lead between the power tranformer and the rectifier string or between the center tap and ground.

If, on the other hand, you get a reading from the negative circuit leg, this is an indication that diodes 3 and/or 4 are defective, C1 is defective, R1 is open or there is a broken conductor within the positive leg. If you get a reading of proper value in the positive leg and none on the negative side, then diodes 1 and/or 2 are defective, C2 is shorted or bad, or R2 is open.

If the voltage readings are significantly higher than nominal, this may be an indication that one or both zener diodes are defective. Be sure of the accuracy of your voltmeter by testing it with a known voltage source. For example, a reading of 15.5 volts on your voltmeter probably indicates that the circuit is working properly and that your voltmeter is off by 0.5 volts. However, a reading of 18 volts indicates a definite problem with the circuit. If you don't know the accuracy of your voltmeter, then it is sometimes difficult to interpret readings.

Using the components specified, you should be able to draw about 125 milliamperes from each of the two output legs. This is for a combined total drain of 250 milliamperes from the transformer. This is well within its limits and may be pushed a bit higher in intermittent applications.

You will find that the dual-polarity power supply is quite useful for a large number of integrated circuit applications. I have used this exact circuit to power slow scan television converters, radio teletype decoders and a large number of other devices which use IC-type operational amplifiers. The 15-volt value was chosen because this is a quite common requirement of operational amplifiers. It has also been found that most non-critical electronic devices designed to operate from 12 Vdc work equally as well from this supply, which provides a voltage approximately 25% higher than required. Be careful here, however. While the higher voltage makes no difference to many types of devices, it will cause many other types to self-destruct.

If you would prefer a 12-volt dual-polarity output, all you need to do is change CR1 and CR2 to 12.6 Vdc units. If they seem to run too hot, increase the value of R1 and R2 to 15 or 18 ohms. Nothing else need be changed, and the resultant output will now be positive and negative 12.6 Vdc with respect to the power supply ground.

A further modification of this circuit would be to make it a dual-polarization, dual-voltage model. A good example of this would be had if you left the positive leg exactly as is and changed the zener diode in the negative leg to a 12.6 volt unit. Now, the supply would provide a positive 15-volt output and a negative 12-volt output. By further modifying the circuit with changes to R1, R2, CR1 and CR2, a wide range of dual-voltage, dual-polarity combinations could be had. Remember, though, as the regulated output voltage differs more and more from the unregulated output voltage (about 18 Vdc), the power dissipation within the zener diodes increases. It is necessary to increase the value of the series resistors to limit this to safe levels.

Some persons have difficulty in understanding the concept of a dual-polarity supply, especially regarding circuit ground. In previous single-polarity supplies, the positive terminal was negative in respect to the negative one. This latter terminal was usually ground. In this dual-polarity supply, the ground is a negative 15 volts in respect to the positive output terminal, but this same ground is also at a potential of plus 15 volts in respect to the negative output leg. Remember, a positive voltage is only positive relative to another point. Likewise, a negative polarity is negative in relation to another point. In a dual-polarity supply, the ground or center tap connection is a common relative point for both the positive and negative supply leads.

PROJECT 7
9-VOLT SERIES-REGULATED SUPPLY

Three or four decades ago, most low-voltage power supplies produced an output of 6 or 12 Vdc. But with the coming of the transistor radio, 9-volt values were used more and more. Small 9-volt batteries were often installed in these radios (and still are).

These are compact, low-current cells which provide good service in low-duty applications. While some 9-volt devices will operate equally well from 6 or 12 volts dc, many others will not and require close tolerance to the 9-volt nominal value.

Inexpensive 9-volt power supplies are available from most hobby stores. These are often self-contained units whose transformers attach directly to the ac outlet through a case-mounted line plug. A two-conductor cable then is run to the battery connection terminal of the radio, and the device operates. These inexpensive power supplies are often called ac battery replacements because they operate from house current. The quality of these supplies varies from adequate to very poor. Often, hum is incurred. This is usually due to lack of regulation. For simple AM portable radios, this may not be a problem; but for sophisticated FM stereo portables, hum can be quite objectionable.

Then, too, there are other devices which require a well-regulated 9 Vdc source. Often, these will not operate properly from inexpensive power supplies designed cheaply and with no electronic regulation.

Series regulation by electronic means is the most efficient, practical form of voltage stabilization available to the average consumer and builder today. Series regulator circuits may be very simple or very complex, depending upon the amount of current which must be passed, the operating voltage, and the degree of regulation. All of them, however, do an excellent job at stabilizing the voltage within very close tolerances, and the output from even a very simple series regulator is far superior to that which can be obtained with most zener diode regulator circuits.

Figure 6-17 shows one of the simplest series regulator circuits available. It offers superlative voltage stability over large swings in current and will safely deliver an output current at a steady 9 Vdc of 400 milliamperes. It is necessary to mount the transistor on a metal heatsink if you intend to draw the maximum 400 milliamperes. If the total current drain will be less than 200 milliamperes, this can be avoided. It's best, however, to go ahead and design for full operating capability, so a heatsink is recommended. This may often be the aluminum chassis or case on which the power supply is built. Alternately, you may purchase a heatsink assembly from your local hobby store, but this will increase the overall price of the project.

Referring to the schematic, it can be seen that the collector and emitter of Q1 is in series with the positive output line. D5 is a 9-volt zener diode. (Actually, a 9.1-volt unit was used when this power

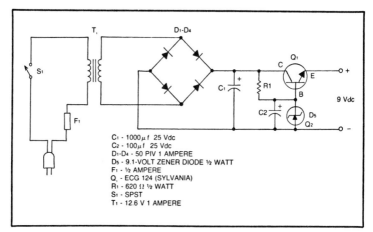

Fig. 6-17. Simple series regulator circuit.

supply was built. This is a common value and will serve just as well as a standard 9-volt unit.) The zener serves as a reference for Q1, which controls the conduction. When the voltage output (unregulated) from the rectifiers is greater than 9 volts, Q1 does not conduct fully. The zener diode sets up the conduction bias for the transistor, which continues to conduct only enough to allow for an output of 9 Vdc.

It can be seen that the power supply is basically a full-wave bridge circuit and is quite common up to the filter capacitor. This is a 1000-microfarad unit, although a 500-microfarad value should serve almost as well. After this point, the series regulator circuitry takes over and is composed of R1, whose sole purpose is to supply the proper operating potential to the base lead of Q1, the series transistor Q1, C2 and D5.

The primary circuit and transformer should first be mounted in a suitable box or compartment. The remaining components will be mounted on a single section of perforated circuit board. Alternately, printed circuit boards may be etched, but for those non-critical applications, a perforated board used with flexible conductors is far more practical as far as the author is concerned. Figure 6-18 shows a suggested layout. C1, the 100-microfarad filter capacitor, is the largest component and will determine the overall size of the circuit board. The remaining components take up little space. If you do not intend to draw the maximum current from this supply, you may wish to mount the series transistor on this board as well. But this project calls for heatsink mounting, so three short wire leads are brought from the circuit board and connected to the transistor.

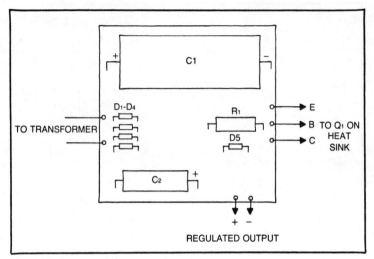

Fig. 6-18. Suggested component layout.

A transistor socket can be purchased from your local hobby store that is designed to be mounted through an aluminum chassis. The metal case of the transistor will be bolted directly to the chassis, but a special insulator which traces the transistor outline is fitted between the two. When installing the transistor, washer and socket, liberal amounts of silicone grease should be applied to the transistor case, the insulator and the chassis. The reason for the insulator or washer is to electrically isolate the transistor from the metal case, which will most likely serve as circuit ground. The collector electrode of the transistor is connected directly to its metal case and must be isolated above ground to prevent a short circuit. The silicone grease assures thermal conductivity. This means that heat will be conducted from the transistor case through the washer and into the larger metal case. The washer/insulator is a good conductor of heat or thermal energy and does not conduct electricity. The same is true of the silicone grease. Once you have your transistor mounted, you may turn full attention to the circuit board again.

Connect each of the three leads to the proper lead on the transistor or terminal on the transistor socket. A reversal here will not only cause the power supply to malfunction, but you can also destroy the transistor. Make sure the emitter lead is connected to the emitter electrode on the transistor, the base lead to the base and the collector to the collector. Double check these connections before applying power. Mount the circuit board near the power transformer by using bolts and nuts, as was previously described. The secondary leads from the power transformer may now be connected to their proper circuit points and the regulated output leads connected to appropriate terminals which will normally be mounted through the power supply chassis or case. A last-minute inspection completes this project and test procedures can now be undertaken.

With a voltmeter properly connected across the output terminals, activate S1 and observe the reading. If nothing happens, check the fuse. If it is blown, this can be a symptom of severe circuit errors or short circuits, especially in the secondary portion of the supply. Examine the primary circuitry and then the secondary. Again, check the connections between the heatsink-mounted transistor and the circuit board. Are the rectifiers connected properly? Are C1 or C2 reverse connected? Has D5 (the zener diode) been inadvertently reversed? Are there any short circuits? If you find a wiring error or short within the regulator circuitry, correct it, replace F1 and try again. If you read 9 volts dc, this is fine; but if you don't, it may mean that one or more solid-state components were damaged by the original short or wiring error. Break the circuit between C1 and R1, and take a measurement between C1 and ground. If you read 18 Vdc (approximately), this is an indication that the basic power supply is functioning properly and the trouble lies within the regulator circuitry proper. Q1 may be damaged. The same applies to the zener diode. Check these two components and replace both if they are defective. You may now reconnect the lead between C1 and R1 and repeat your test procedures.

You can see that it is absolutely necessary to assure that proper wiring has been completed within every portion of the power supply before activation. With the basic power supply circuits discussed earlier, about the only components that could be destroyed by an inadvertent short circuit or wiring error were the inexpensive silicon rectifiers. But as the power supply circuits become more complex, expensive components enter the picture, and more caution is necessary in wiring and soldering to prevent costly replacements.

If during the test procedure you get an erratic reading, one that fluctuates continuously or is too high or too low, there is a good chance that the zener diode is of the improper value or is defective. This assumes that the output voltage from the rectifiers and C1 seems normal. Erratic output can also be attributed to a faulty power transformer, intermittent series transistor operation and a number of other extremely rare occurrences. We are dealing with worst case analysis here in an attempt to take the reader through every possible malfunction. If you observe proper building practices, take your time and perform the needed checks *before* initial activation, your circuits should work properly the first time and every time.

Once you have checked your power supply thoroughly and found it to be operating properly, you may attach the output terminals to any electronic device requiring no more than 400 milliamperes of operating current. The regulation efficiency of this supply is excellent and it should satisfy the needs of even highly critical circuits.

If you intend to operate transistor radios, tape recorders, etc., you may wish to replace the output terminals with a standard 9-volt battery clip which is a terminal strip with output leads identical to those of a standard 9-volt battery. Using this arrangement, any device which operates from such a battery may be connected in moments to this power supply. Alternately, you may wish to install a two-conductor cable at the power supply output and fit the opposite end with one of the many different types of power plugs used to mate with radios, tape recorders, etc. The output arrangement which you ultimately decide to use will depend upon individual preference and intended applications. Regardless of which output method you use, be careful to guard against accidentally shorting the positive and negative terminals. Such an occurrence may only cause the line fuse to blow, but more than likely, you will also take out the series transistor which, next to the power transformer, is the most expensive single component. This is one of the drawbacks of simple series regulator circuits. They are quite susceptible to damage from short circuits and high current demands which fall outside of their maximum ratings.

Should you notice that the series transistor is operating at an extremely high temperature or if device failure occurs quite frequently, you may be drawing too much current from the supply. If this is not the case, the surface area of the heatsink is probably not adequate to keep the transistor operating at a safe temperature. Here, it will be necessary to provide a larger metal surface or to attach a finned heatsink to the device. This assumes that the transistor has been mounted properly and that a good thermal connection has been established between the device body and the heatsink. If you have failed to use the silicone grease, very little of the heat built up in the device will be transferred to the sink, and rapid failure will occur, especially when operating at near maximum ratings.

This power supply lends itself to some modifications. If you replace D5 with a lower value zener, the output can be dropped to 5, 6, or 7½ Vdc. You could probably also obtain a 12-volt output if current demands are not too high, although this would be pushing the upper limits for good regulation. The series transistor must have considerably more voltage than is required at the output to act upon or regulation is hampered or non-existent. For best results with a 12½-volt output, a transformer of a higher secondary voltage (say 20 Vac) would be preferred.

While the current rating at the output of this supply is adequate for most small electronic devices, especially those requiring a 9-volt operating potential, some situations may require a higher current capability. This could easily be had by substituting the present power transformer with one with a larger power rating. Of course, you would have to increase the forward current rating of D1 through D4 as well. Every other component could remain the same, with the exception of Q1, which would have to be replaced with an NPN transistor capable of handling the higher output current. Searching through any manufacturer's cross reference catalog of solid-state components will reveal a number of devices which will be adequate. Since we used a Sylvania transistor in this circuit, researching the Sylvania catalog would indicate that an ECG128 could probably serve as a direct replacement and could safely withstand an output current rating of a full 1000 milliamperes, or 1 ampere. It would be necessary to install this device in a larger heatsink, but everything else could probably remain the same. Certainly, other transistors would serve equally as well, although it might be necessary to make some minor circuit changes. A Sylvania series transistor was specified here simply because this was the one which was used when the supply was originally built. Every other manufacturer of transistors offers an exact replacement for this device as well. It is not the purpose of this book to specify one manufacturer's products over another in such a manner as to make it seem that one type is to be preferred. However, it is a good idea to stick with the components of well-known manufacturers when maximum ratings will be depended upon. While this is not often true, the author has occasionally run into problems with a few questionable brands that were brought through surplus channels. Nine times out of ten, surplus transistors and other solid-state devices will work equally as well as newly boxed products from a manufacturer. (Indeed, most of the surplus solid-state components have probably been made by these companies.) Occasionally, problems can develop, especially if surplus transistors were discarded by the manufacturers because they weren't quite up to specifications. Again, this most often applies to devices which are operated at their maximum ratings.

PROJECT 8
POWER SUPPLY /BATTERY CHARGER

Today, it is quite common for many types of electronic devices to operate from rechargeable batteries. These may range from small 1.3 volt cells to large 12-volt battery packs. While being quite expensive, the savings are had over a long period of time when the exhausted batteries are given a fresh charge again and again. What is a battery charger? It is simply a dc power supply. When attached to a battery, the current from the power supply is forced through the internal plates in the reverse direction from the current flow when the battery is supplying power. This creates an electrochemical reaction and the charge is restored.

Battery chargers can be highly complex power supplies with automatic current and voltage reduction built in, constant trickle charging, high, medium and low charge rates and automatic short-circuit protection. They can also be as simple as the circuit shown in Fig. 6-19. This may look similar to our series-regulated supply discussed previously and works in very much the same manner. The only difference is that the regulation is manual rather than electronic and automatic. By adjusting the value of R1, the output voltage changes. This is what the series regulator circuit in the last project did, in a manner of speaking, except the changing of the internal conduction of the transistor was handled by electronic means rather than by a human thumb and forefinger on a potentiometer.

For our battery charging purposes, however, this type of supply works far better and is simpler and less expensive to boot. Referring to the schematic, we see our old standard primary circuit, followed by a full-wave bridge rectifier and filter capacitor. At this point, we have established a dc output which has been filtered and can be measured at somewhere in the neighborhood of 38 volts dc.

By adding R1 and Q1, it is possible to adjust the final dc voltage output to any point between 0 and approximately 30 volts. Q1 is quite overrated for this application, but was chosen because it will not need a heatsink. It is capable of passing about 4 amperes at maximum rating. However, as part of this circuit, it will pass no more than 300 milliamperes. It is safe to draw up to 600 milliam-

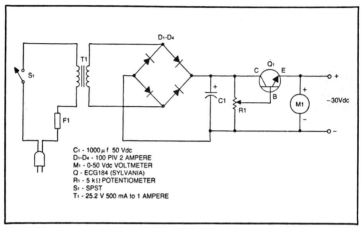

Fig. 6-19. Power supply/battery charger circuit.

peres if the output voltage is above 15 Vdc. As the voltage output is lowered, the non-conducting load on the transistor becomes greater and current demand must be decreased.

M1 is a 0-to-50 Vdc voltmeter and may be purchased through surplus channels and at some electronic hobby stores. This allows the precise setting of an output voltage level to a battery which is under charge or to an electronic circuit. Regulation here is not very good and the output voltage will fluctuate with varying demands. The main advantage of this supply is found in the fact that it can safely power low current loads requiring anywhere from 1.5 to 30 Vdc. From a practical standpoint, however, it is difficult to easily achieve any value of less than 6 Vdc with any accuracy. Perhaps a 0 to 10 Vdc meter would help in setting a low operating value, but might be damaged if you exceed this output.

Construction of this project is exactly like that of the simple full-wave bridge power supply project discussed earlier, and indeed, many of the components are similar. When I built this supply, I used a full-wave bridge IC package, soldering the appropriate leads directly to those of the power transformer. I then attached the small case to the transformer shell by a drop or two of epoxy cement. A hole is drilled in the power supply case to accept the potentiometer R1. Leads are brought off this device for attachment to the back of the rectifier stack, to the transistor base and to the positive terminal of C1. Another hole is drilled to mount a five-contact terminal strip. Only three of the contacts are to be used, with the center or ground terminal serving as the attachment point. Figure 6-20 shows how the transistor is connected and also includes the connecting points

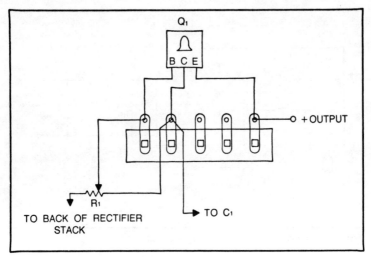

Fig. 6-20. Transistor wiring diagram.

for the positive output lead, the R1 leads and the connection to C1. The reason that a piece of perforated circuit board was not used was due to space limitations during the author's construction project. Certainly, a small section of perforated circuit board could serve as the mounting platform for the diodes (if discrete units are to be used) and the transistor, as shown in Fig. 6-21. Of course, it would still be necessary to mount R1 through the chassis wall for most practical operation.

Another hole is cut in the chassis, through which M1, the voltmeter, is mounted. If you look hard enough, you will probably be able to find a miniature meter which will require a 2-inch diameter hole for mounting. This hole may be cut with a circular die. Alternately, holes may be drilled through the case in a circular pattern and tin snips used to remove the sections in between. When a jagged

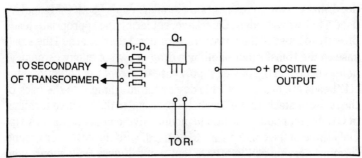

Fig. 6-21. Component mounting on perforated circuit board.

hole is finally formed, a small file will smooth out the edges. Once the meter is installed, the indicator facing should hide most of the irregularities and the final appearance should be that of a professional installation.

I used an aluminum Bud Box™ for construction of this battery charger/power supply. This made a convenient hand-held unit that was quite easy to pick up for accurate meter observation. The arrangement of the through-the-wall components is shown in Fig. 6-22. You may elect to design your project a bit differently, and this is encouraged. As the author built his to satisfy an individual preference and operating intention, you should do the same with yours.

As always, be careful with the polarized components. These include the rectifiers, filter capacitor, Q1 and M1. A reversal of M1 could cause the needle indicator to move downward, and a reversal of the leads is all that is required. Recheck the connections to Q1, making sure that they have not been reversed. If you use the ECG 184 specified, the emitter lead will be on the far left, the collector in the center and the base on the right, as viewed from the front. Do not allow the metal strip contact on the back of the package to come in contact with the chassis, as this is also connected to the collector electrode. If there is any chance of this, wrap the transistor body with insulating tape.

If you have connected your meter properly, you won't need a separate voltmeter to test this circuit, as was necessary for all the

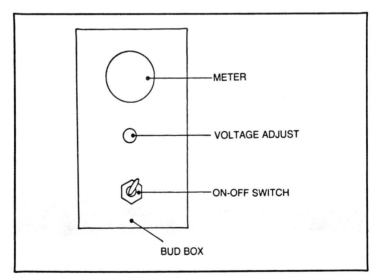

Fig. 6-22. Arrangement of through-the-wall components.

173

previous projects. Simply activate S1 and watch for an indication on the meter. If you get nothing, check the primary circuit for a blown fuse, lack of ac potential, etc. Assuming that everything is all right here, look for shorts and polarity reversals in the secondary. If you suspect that the meter is malfunctioning, temporarily remove it from the circuit and check with an outboard voltmeter.

Once you have obtained output voltage from the supply, vary R1 and note the change in meter reading. This may best be done by connecting the output terminals to a load of some sort, as the current drain will have an effect on the output voltage and R1's ability to control it. If you are able to vary the output from near 0 to roughly 30 volts (possibly higher), then everything is working satisfactorily.

To charge small batteries or to give a trickle charge to 12-volt automotive batteries, simply attach the positive power supply lead to the positive battery terminal and do the same with the negative lead and terminal. With R1 in a low-voltage position, activate the power supply by tripping S1, and rotate R1 until the meter reads the same as the intended battery voltage. That's all there is to it. You can give your small rechargeable batteries a full charge in a few hours or overnight (depending upon the size), and you can keep automotive batteries fully charged by supplying a constant trickle of current.

This device makes an excellent all-purpose shop power supply. It is extremely versatile because of its ability to be manually adjusted (regarding voltage) to variable loads. If you concentrate on compactness, it can be carried in a coat pocket wherever you go.

As was mentioned earlier, the regulation of this power supply is not adequate to properly power complex electronic circuits which require excellent voltage stability. If better regulation is needed, you can connect the outboard zener diode regulator to its output, as was described in an earlier project.

This circuit, however, is not really designed for highly critical applications, but it is an excellent little device which will find many uses around the electronics shop and in many other areas of the home.

PROJECT 9
VERSATILE VOLTAGE DOUBLER SUPPLY

If you do a lot of electronic experimenting, sooner or later, you will find yourself in a position where it is necessary to build a dc power supply from available components. This happens quite often when a new project is built and there is no present power supply in your possession which will provide the voltage and current necessary for operation. Also, emergency ac-derived power supplies are sometimes whipped up on the spur of the moment when batteries fail or even when another dc power supply has become hopelessly damaged.

In situations such as these, it is usually necessary for you to be extremely versatile in your thinking, because components often are not as versatile. However, with a little bit of forethought and some basic knowledge in electronics (which has already been provided in earlier chapters of this book), you may find that power transformers are extremely adaptable devices and are not limited to just a small range of output voltages when used in dc power supply construction.

For example, let's assume you need a dc power supply with an output of somewhere between 12 and 18 volts. But all you have on hand is a transformer with a 6.3 volt secondary. Obviously, none of the standard rectifier configurations will provide enough dc voltage output. About the most you would get from any one of these would be about 9 volts.

Believe it or not, a transformer with a 6.3 volt secondary can be made to deliver a wide range of dc voltage outputs. As a matter of fact, this transformer could theoretically deliver thousands of volts, but the rectifier circuitry would be quite complex, extremely large and totally efficient.

What we're talking about here is voltage multiplication. This was discussed in an earlier chapter and is quite helpful in power supply design, as long as multiplication factors do not get too high. The voltage doubler is the most popular form of voltage multiplying circuit and is quite commonly used in high voltage power supplies. With this rectifier configuration, a 1000-volt transformer secondary can easily deliver 2500-2800 volts dc at good efficiency. To recap a bit, a voltage doubler circuit delivers twice the peak ac voltage. A transformer with a 6-volt secondary has a peak voltage rating of approximately 9 volts. Therefore, a voltage doubler will deliver twice this value. An easier way of calculating dc output (under light to moderate loading) from a voltage doubler circuit is to say that the final dc output will be equal to 2.8 times the secondary RMS voltage.

In the case of our 6.3 volt transformer, a voltage doubler circuit will deliver a dc output of 2.8 times 6.3 volts, or 17.64 volts. When current is drawn, this value will typically drop to around 15 volts dc. Knowing this, our problem is solved, in that we can obtain the needed output (12 to 18 volts) from the 6.3 volt transformer. Under heavy loading, the output voltage will be very close to our 12-volt needed minimum. Alternately, a simple voltage regulator circuit can be added to produce any voltage within a range of 6 to about 16 Vdc.

Figure 6-23 shows the simple full-wave voltage doubler circuit. Notice that only two rectifiers are used, along with two capacitors which are wired in series with each other. The voltage doubler circuit shown here is actually two half-wave rectifier power supplies combined in series-aiding. One diode and capacitor combination works from the positive portion of the ac sine wave, while the remaining combination works from the negative. It is necessary that C1 and C2 be of the same value in order to equalize rectification from both halves of the cycle. Ideally, C1 and C2 will be identical capacitors made by the same manufacturer.

You can construct the entire circuit on a piece of vector board. A suggested layout is shown in Fig. 6-24. Be sure to observe capacitor polarity, in that a reversal will cause inoperation. Also, note that the bottom of C1 (negative) and the top of C2 (positive) are connected directly to the transformer secondary lead. Make certain that the polarities of D1 and D2 are arranged as shown in the schematic drawing.

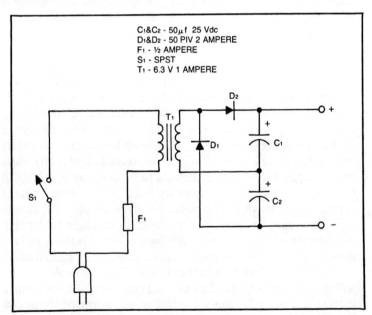

Fig. 6-23. Full-wave voltage doubler circuit.

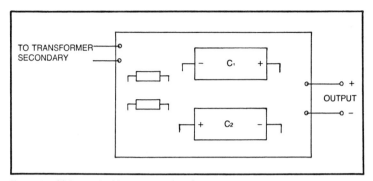

Fig. 6-24. Voltage doubler components layout.

The primary wiring circuit is straightforward and is protected by a one-half ampere fuse. This is adequate protection for a power supply using the components shown in the schematic. By removing D1 and C2 from the schematic diagram, it can be seen that a simple half-wave supply is formed. By removing D2 and C1, another half-wave supply is left. An earlier power supply provided two voltage legs from a single transformer, one positive and one negative. This design also pulls two voltage legs from a single secondary, but then combines them in series-aiding fashion. Thus, we have doubled the normal output.

Install the transformer and primary circuitry in a small plastic or aluminum box. The wired circuit board should be mounted in close proximity to the transformer secondary winding leads. Four small bolts can serve as mounting legs, or you may simply decide to wrap the entire circuit board in insulating tape and attach it to the chassis using any of a number of bonding compounds. (Note: If you decide to mount the circuit board using the latter method, it's best to make sure the circuit is operational first.)

Closely inspect all of your wiring and solder connections. If you have used an aluminum box, it has probably been necessary to drill holes through the surface for installation of the switch, the fuse and to make an exit for the line cord. Examine the interior of this box closely to make certain that no small shards of aluminum have been allowed to remain. These can eventually wind up in the electronic circuitry and cause malfunction as well as component damage.

If all seems to be in order, make certain S1 is in the off position and insert the line plug in the wall outlet. With a voltmeter across the power supply output, activate the switch and note the reading. You should get an indication of a little over 17 volts. If not, make the usual inspection, looking for reversed polarities, broken conductors, shorted conductors, etc. You may also wish to run an additional check relevant to this particular supply. With the circuit activated, take a voltage reading across the two sides of C1. If this leg of the circuit is functioning properly, an indication of 8 to 9 volts will be had. If you get nothing here, check the reading across the leads of C2. Again, a reading of 8 to 9 volts indicates that this half of the supply is operational. If either leg is malfunctioning, no output will be had at the power supply terminal. A malfunctioning leg can most often be attributed to a defective diode or capacitor, assuming that the circuit is otherwise in a good state of repair.

This supply was designed for general purposes. Voltage doubler circuits require good regulation and to do this, at least twice the capacitance is needed. This is not technically accurate but is a statement of practicality. Notice that in this power supply, we have used two 500-microfarad capacitors. This is not equal to a 1000-microfarad capacitor in a power supply which uses a standard rectifier configuration. As a matter of fact, due to the series nature of the circuit, capacitors C1 and C2 combine for a total output capacitance of about 250 microfarads. For better regulation, it might be desirable to replace C1 and C2 with 1000-microfarad units. This was not done for this particular circuit, because if this degree of regulation is desired, a simple electronic regulator circuit would be the less expensive route to take. Any of the previously discussed regulator circuits will work as well with this design as any other power supply.

If you would like to make a simple modification to this supply which simply involves soldering a short length of conductor to the positive lead of C2, this power supply becomes even more versatile. Figure 6-25 shows the output section of this supply with the attached conductor. This has been labeled B for clarification purposes, while the two former leads are labeled A and C. What we end up with is a full-wave voltage doubler supply that will deliver a maximum of approximately 18 volts *and* two half-wave supplies that will deliver maximum outputs of 9 Vdc. The latter arrangement is bipolar with respect to circuit ground, which is lead C. The A output is positive to B, while the C output is negative to B. Look what we've ended up with—a bipolar power supply and a full-wave single-voltage power supply, all from one transformer and at the expense of only two diodes and a like number of capacitors.

Remember, while the regulation of the 18-volt output is good for basic power supply, the regulation of each 9-volt output is

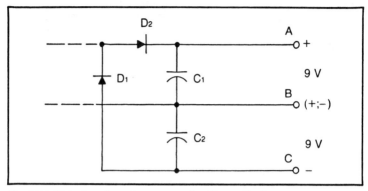

Fig. 6-25. Output section of power supply with modification.

nothing to brag about, since both are derived from basic half-wave supplies. Also, it is necessary to consider current drain. A 1-ampere secondary is all that is available to us using the transformer specified. Therefore, current drain from either 9-volt leg must be kept below this level. (Figure 750 to 800 milliamperes to be on the safe side.) If you use both supplies simultaneously, the current drain adds, so you are limited to a maximum of about 400 milliamperes from one supply and the same from the other. Now, if you use the full-wave 18-volt output, you are still limited. No, you cannot draw 800 milliamperes at 18 volts safely. Since the voltage has been doubled, the current must be cut in half. The maximum safe range from this circuit using the components specified is 400 milliamperes at 18 Vdc.

Do you wonder why you can't draw 1 ampere of current at the higher voltage, since the transformer is rated to deliver this amount of current? The reason is that we have pulled two power supply legs off of a single secondary. If 1 ampere of current were drawn from the voltage doubler output, the ac current drain from the transformer secondary would be 2 amperes or more. This obviously exceeds the transformer's rating and damage would quickly result. Of course, there is nothing to stop you from using a transformer with a higher power rating, but the idea of this project is to show how nearly any reasonable voltage can be obtained from a number of different filament transformers. The same principle can be applied to transformers with any number of different secondary voltage ratings. Regardless of the secondary winding value, a voltage doubler circuit will always produce a no-load output voltage of approximately 2.8 times the RMS reading.

PROJECT 10
FULL-WAVE VOLTAGE TRIPLER SUPPLY

The previous voltage doubler circuit was designed to fulfill one of my specific needs over ten years ago. If I had not been familiar with voltage multiplication, the power supply would not have been possible, at least not to me and on the spur of the moment. Many persons are intimately familiar with the simple full-wave voltage doubler circuit. In most ways, it is no more complicated than a full-wave center-tapped configuration and even simpler than a bridge circuit.

But past the doubling stage, many experienced electronic experimenters and technicians are totally lost. They can sketch from memory a voltage doubler but know nothing about voltage triplers, quadruplers, quintuplers, etc. Actually past the quadrupling stage, most voltage multiplier circuits become some complicated and/or inefficient that they are not applicable to many devices. Voltage triplers, however, can certainly be made useful, especially when a specific dc output needs to be obtained from a transformer which will not deliver this value using standard rectifier configurations.

Figure 6-26 shows a voltage tripler circuit of full-wave design which uses a 12.6-volt, 1-ampere filament transformer to deliver a dc output of a little over 50 volts. Just like the doubler circuit, the tripler multiplies the peak value of the secondary ac by a certain factor (in this case, three). An easier way to judge, though, is to say

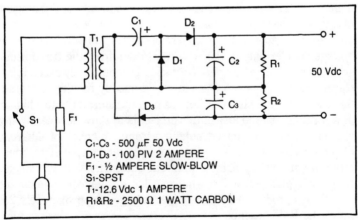

Fig. 6-26. Full-wave voltage tripler circuit.

180

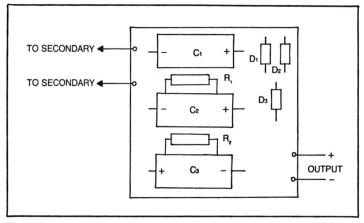

Fig. 6-27. Voltage tripler component layout.

that a voltage tripler will increase the secondary RMS voltage rating of a transformer by 4.2.

Notice that the voltage tripler circuit shown schematically uses three diodes and three capacitors. Voltage triplers, unlike doubler circuits, are inherently unbalanced and are more difficult to filter because there is a 60-Hertz ripple frequency involved, along with one at 120 Hertz. Higher capacitance will be required than with a doubler or quadrupler circuit for similar filtering efficiency. Notice that this power supply contains two resistors at the output. This is a bleeder string consisting of R1 and R2. Together, they form a 5000-ohm bleeder resistor and could be replaced with a single unit of this value. However, dividing the total bleeder resistance by two units and placing each across the two output capacitors, a bit of equalization is added. It is necessary that the two capacitors be identical to provide a reasonably balanced output, and the two added resistances tend to equalize internal resistance within C2 and C3. This is far more important when capacitors are wired in a series string than in this particular circuit, but the principle is the same.

A larger section of perforated circuit board will be required to contain all of the components of the secondary circuit. The three capacitors will require the largest space. The component layout is totally non-critical and can be arranged to suit the builder's desires. Figure 6-27 shows the layout I used, which was designed to contain the components specified in the parts list. If you elect to use larger value capacitors, you may have to increase the size of the circuit board. One note: The voltage rating of all capacitors in this circuit may be slightly reduced. 35 Vdc capacitors will provide an adequate safety margin, but 50 Vdc units (which are specified) may be as easily available and will provide an even higher degree of reliability. It is necessary that C2 and C3 be rated for at least 35 Vdc, but C1 can be safely reduced to 16 volts. However, if two capacitors are of one voltage rating and a third is at another, certain capacitance mismatches may occur between the three, because two will obviously be of one design and the third of another. In voltage multiplier circuits, it is best that all capacitors be identical and from the same manufacturer.

This power supply is a bit more complex than any of the previous ones. It is quite easy to reverse polarities at any of the six capacitor connection points. Reversing a diode in such a circuit is not exactly unheard of either. When wiring the circuit board, take the extra time required to see that all components are properly laid out and connected as shown in the schematic drawing. A reversal here will cause your power supply to malfunction and diodes and even capacitors may be destroyed. Take each connection as a separate maneuver and compare what you have on your circuit board to what is contained in the schematic drawing. Once you are sure everything is in order, make your solder connections. A reexamination is in order at this point.

Checkout procedure is quite simple after the components have been mounted in a protective aluminum or plastic case. After the line plug has been inserted into the wall outlet, activate S1 and note the reading at the output terminal on an accurate dc voltmeter set to a range which will indicate at least 55 volts. Proper operation is indicated when a reading of about 52 volts is obtained. Please do not take the voltage figures I give as exact. Depending upon your line voltage and a number of other factors, the proper output voltage reading for this and most other basic supplies in this book which do not use electronic regulation can vary by 20% in some instances. Suffice it to say that if you are somewhere in the 50-volt range, your power supply is okay. If, however, you get an indication of 10 or 20 volts, obviously something is wrong.

Should your power supply not produce an acceptable output, check the primary circuit to make certain that voltage is being delivered to the primary windings of T1. If all appears to be in order here, examine the secondary circuit for a possible short circuit, a broken conductor or a reversed component. If you do find that you have messed up the polarity of one or more secondary circuit components, it may be necessary to test all of them, especially the rectifiers, to see if any damage was incurred by the wiring error.

If, on the other hand, you find no apparent errors in the secondary wiring, run a voltage check on each of the three circuit legs with a dc voltmeter. Assuming that you have a secondary ac output, you should read approximately 17 volts across each of the three filter capacitors. Make sure you observe voltmeter probe polarity when taking these readings and be careful to stay away from the ac primary line. When you find the leg or legs which are not producing an output, you can be certain that the problem lies with a defective component, short circuit, open circuit or a wiring error in this leg(s) alone.

Once the supply is completed and known to be operational, you now have an ac-derived 50 Vdc source which has been obtained from a 12.6 volt filament transformer. A 1-ampere unit was specified here, so a safe current drain from the dc output of the tripler would be ⅓ this value, less about 20% for circuit losses. If the power supply is to be used intermittently, you can probably disregard the 20 percent derating, but check the transformer after a normal period of operation to make certain it is not overheating. Remember, when you draw approximately 330 milliamperes from the output of this power supply, the total drain on the transformer secondary is about 1 ampere, its maximum rating.

PROJECT 11
SWITCHABLE FULL/HALF-VOLTAGE POWER SUPPLY

One of my favorite power supply designs is one I discovered as a teenage electronics experimenter. The design is not seen too often but is very handy for a number of power supply applications, especially for radio frequency transmitting applications. This design can be built into any power supply and mainly concerns the primary windings and associated circuitry. Figure 6-28 shows the schematic drawing. The transformer secondary windings and the remainder of the secondary circuitry are not given any values, simple because this idea will work with any ac-derived power supply. The idea here is to build a power supply which will operate from 230 volts ac and produce a specific dc voltage output at this level. Then, by switching the primary voltage to 115 Vac, the output voltage from the supply is cut in half.

To do this, use two power transformers. The primary windings are connected in series, while the secondary windings are connected in parallel. This means that the two transformers may now

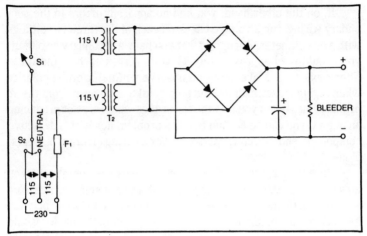

Fig. 6-28. Full / half-voltage power supply circuit.

be powered by a 230-volt ac source due to the series connection and the primary windings (in total) will still deliver the rated voltage of a single primary, but twice the amperage.

It is recommended that both transformers be identical. Connect the primary windings exactly as shown. A reversal here can cause the entire supply to malfunction. Also, connect the secondary windings in parallel-aiding. This means that you connect the top winding lead from one transformer to the top winding lead of the other, and so forth. If you reverse these leads, the transformers will be connected in parallel-opposing, and there will be no output. Once the series connections are made, you can then connect the secondaries in parallel and take an ac voltage measurement at the combined secondary output to make certain your connections are correct. The ac output voltage from the secondary winding combination should be exactly the same as from a single winding.

Again, the secondary circuitry from this point on can be anything you make it. The schematic drawing shows a full-wave bridge rectifier, but you could as easily use a doubler, half-wave, full-wave center-tapped, or any other configuration. You can also add regulating circuits if desired, but remember that they will not be operable when the supply is switched to half voltage.

The primary supply voltage is delivered by a three-wire 230-Vac circuit. This is the same type of supply line which delivers power to your electric stove, dryer and other heavy-duty appliances. As is shown by the schematic drawing, there are two hot wires and a neutral. There is a 230-volt potential between the two hot wires and 115 volts from either wire to neutral. S2 is a single-pole double-throw switch which will select either the left-hand hot wire or the central neutral conductor. The combined transformers actually form a single transformer circuit which has a 230-volt primary winding. When S2 is in the left-hand position in contact with the hot wire, the primary receives full voltage. But when S2 is switched to the neutral position, 115 volts is delivered to the 230-volt primary, and the secondary output will also be reduced by half. This means that if the secondary windings produced 100 Vac at 230 volts, the output would be 50 volts at a primary drive of 115 Vac. Incidentally, this schematic drawing shows the secondary windings connected in parallel. They can also be connected in series, as with the primary windings, and the total secondary output will be twice the voltage value of a single transformer secondary. Using the above example, if a parallel connection of the secondary windings produces a 100-Vac output with a full 230-Vac drive, then series connection of the secondary windings will produce a 200-Vac output, again with full drive. When the primary voltage is half, the secondary voltage will still do likewise.

This type of power supply will often supply 800 or more volts in the high voltage position and is used to supply operating voltage and current to a radio frequency amplifier. When it is desirous to decrease power output of the rf amplifier, this is accomplished by halving the voltage to the amplifying stage.

For experimental purposes, a power supply which will operate from 230 or 115 volts is quite desirable, and two output voltages are useful for many applications. While two transformers were used in the same circuit, you could as easily use a single transformer which was equipped with a 230-volt primary winding. These devices are not as common in hobbyist circles, especially where smaller transformers are concerned. Most types you are likely to run into will have standard 115-volt windings at the primary.

Since specific components have not been listed in this project, it is necessary to discuss ratings, especially regarding the switches in the primary circuit. The amount of current which is drawn through the primary windings will be determined by the size of the transformers and, of course, the power demand from the dc output. Use the Ohm's Law power formula of $P = IE$ to determine the total power drain from the dc output. For example, if your power supply delivers 1000 volts dc and the current drain is 500 milliamperes, then the dc power drain is 1000 × .500, or 500 watts. Of course, there are certain inefficiencies within any electronic circuit, so the total drain from the ac power line will be slightly higher (a maximum of 10%). At 230 Vac, a power consumption of 550 watts would represent a current flow of about 2.5 amperes. This means that S1 and S2 should be rated to handle at least 3 amperes and, preferably, 5 amperes to provide an adequate margin. F1 should be rated at 5 to 6 amperes (slightly more if the power supply is to be subjected to high, instantaneous surges). Just make certain that all components within the primary winding are rated to handle the current drain.

Now, all of our calculations so far have dealt with the supply operating at 230 Vac, the high voltage mode. When the supply voltage is dropped to 115 volts, not only is the output voltage value at the secondary windings halved, but so is the transformer power rating. If the transformer will safely deliver 550 watts maximum at 230 Vac, then the best it can do at 115 Vac is 275 watts. Why is this? Because of the current through the windings. In a series circuit, which is what we have at the primary windings, current is the same at all points. Using the previous example, 2½ amperes was the maximum current rating for these windings. 2½ amperes at 230 Vac means a total power of about 550 watts. But at the lower voltage, the current ratings of the windings remain the same. The maximum current through these windings cannot exceed 2½ amperes. Using Ohm's Law power formula ($P = IE$), 2½ amperes at 115 volts is approximately 275 watts. In other words, you can halve the voltage but you can't double the current. Ratings will be exceeded. The rating discussed here is not exactly power; rather, it is the amount of current which the size of conductor used in the primary windings can safely withstand without becoming overheated.

Using this circuit with low voltage transformers, you can build a very simple power supply which offers a 12-volt dc output in one mode and 6 volts at the flip of a switch. With other transformers, you might even have 5000 volts in the high position and 2500 at half power. The combinations are as endless as the number of different transformers which can be substituted in this circuit. Incidentally, if you ever get tired of the two-voltage feature and would prefer to return to standard 115-Vac operation, you can simply wire the transformer *primaries* in parallel and obtain maximum voltage from 115 volt operation. Since two transformers are used and their primaries are in parallel, the total primary circuit may draw twice the maximum current rating of a single transformer. This was not true with the previous series connections.

PROJECT 12
MULTI-OUTPUT, ADD-ON REGULATOR

Many inexpensive power supplies which can be purchased through electronic stores and hobby outlets do not provide for electronic regulation. This is fine for electromechanical devices such as relays, motors and for a few simple types of electronic circuits. However, many different kinds of electronic devices require considerably better regulation for proper performance than can be provided by these inexpensive power supplies.

An earlier project was designed to be added onto an unregulated supply to provide a regulated dc output. However, this particular project was designed for a specific output voltage. In many instances, different voltages will be required, especially when one inexpensive power supply is to be used for test purposes or to power a large number of different electronic circuits.

Figure 6-29 shows an add-on regulator circuit which is designed to produce a number of different output voltages, depending upon the switch position of S1. This selects any one of four zener diodes, each rated at a different voltage. The values of these diodes are not specified in this project and will be selected by the builder to best fulfill his individual needs.

The circuit is identical to previous zener diode regulator designs. A switch has been added for selection of different output values, but only one diode is in the circuit at any one time. The

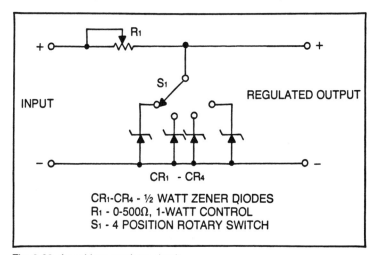

Fig. 6-29. An add-on regulator circuit.

series resistance, R1, is made variable to optimize power supply performance. This resistor might be replaced with a number of fixed components which are selected by another rotary switch. A linear taper variable control is used in this particular project because of the versatility of operation. The value specified is 0 to 500 ohms, but this will vary depending upon the input voltage and the amount of current to be drawn from the supply. Depending upon your application, you may need to increase the value to a maximum of 1000 ohms or more. If you can find a variable control which is rated at a higher power level (say, 5 or 10 watts), this is to be preferred, since it will allow higher input voltages to be used safely to produce low outputs. These high-wattage controls can be quite expensive when purchased new, but can be had from surplus outlets for less than five dollars. While not rare, these variable resistors are sometimes difficult to locate in certain areas, so the common 1-watt control is listed for this project. It should be remembered, however, that if you have a high input voltage to the regulator circuit and have the switch placed across a low-voltage zener (3 or 5 volts), a large amount of voltage will have to be dropped by the series resistor and it may exceed its power rating.

The entire project will easily fit into a small plastic or aluminum box. You may use perforated circuit board as a mounting platform for the zener diodes, but you may find it more desirable to use S1, the four-position rotary switch, for this purpose. It is a simple matter to clip the zener diode leads short and use the switching contacts for mounting pins. Incidentally, if a 6, 8, 10 or more position rotary switch is available, you may elect to use this, along with a comparable number of zener diodes, to further expand the output flexibility of this add-on regulator.

It will be necessary to drill two holes through the metal or plastic container for mounting R1 and S1. Input and output terminals may be of the screw-on variety, although you may elect to use two conductor cables, terminal strips, or any other input/output coupling method which is most suited to your particular requirements.

There is very little complexity in building this circuit. Just be certain the polarity of the zener diodes is observed at all times and that you properly identify the positive and negative input/output terminals. That's all there is to it. This project can often be correctly completed in less than an hour's time. This will, of course, depend upon the experience of the builder. If you are new at building your own electronic circuits, don't rush. An extra fifteen or twenty minutes devoted to meticulous construction can mean a savings of hours in troubleshooting a defective circuit. It can also mean a savings in dollars if rushing causes component damage during initial tryout.

Before testing this circuit, it is necessary to explain that while it is possible to obtain a great many different voltage values from this circuit, a lot will depend upon the input voltage value. As a general rule of thumb, the input value should be no more than 2 or 3 volts higher than the desired output at low voltage ranges (3 to 9 volts) and no more than 5 or 6 volts higher for outputs of 12 volts and above. This is to avoid exceeding the operating maximums of R1 and the zener diodes. For example, if you desire a 3-volt output, the input voltage should be no higher than 5 or 6 Vdc. If the input voltage were excessively high (25 Vdc or more), you would quickly destroy the series resistor.

Testing is very simple. Connect a filtered dc input voltage to the input terminals and place S1 in the correct position for the desired output voltage. With voltmeter probes across the output terminals, observe the indicated reading. This should be equivalent to the zener diode value. It may be necessary to adjust R1 to obtain the correct value. It is best to start with the shaft in the mid-range position and then adjust for increasingly higher values of resistance. You will reach a point when the output voltage will drop below the desired value. Back off on the resistance until the proper output returns and then back off a little more. This should be the ideal operating position, although further adjustments may be necessary when the regulator is operated under load.

This can be thought of as a low-current regulator, although in certain applications (and with some circuit modifications), high amounts of current can be drawn. For most purposes, it is best to keep the current drain to 50 milliamperes or less. This is especially true when it is necessary to use a high resistance setting of R1. When R1 is at the center of its maximum value or lower, a bit more current may be drawn. If R1 is replaced with a 5 or 10 watt unit, values of 100 milliamperes and more may be drawn in the high resistance position.

By incorporating zener diodes of the correct values, you can tailor this add-on regulator to suit your individual operating needs. To make the circuit even more flexible, you might leave a blank position in component S1 for the external insertion of a zener diode of a different value than those wired into the circuit. To do this, simply attach short lengths of conductors to the contact points where the zener would normally be installed. Bring these conductors out to a screw-in terminal strip or plug which will accept the leads of a zener diode. This allows you to have access to the internal circuitry from the outside and you simply bend the leads of the zener and insert them into the plug or terminal. With the switch in the auxiliary position, this externally inserted diode is made a part of the regulator circuit. The addition of this modification will allow you to increase the number of output voltages available without having to resort to a rotary switch with a greater number of contact points.

It may also be desirable to add a faceplate to the series resistor. Its shaft can be fitted with an indicator knob and the faceplate scaled. This combination will allow you to set the correct resistance reading for a specific output voltage, mark it on the scale, and return to it any time you desire to use this output. Once the initial testing and scale marking are completed, it will not be necessary to use an external voltmeter to assure most efficient operation.

I think you'll find this circuit to be quite an addition to your workbench and it allows you to use any inexpensive nonregulated supply to produce a regulated output. Even simple half-wave power supplied will do a reasonably good job, when used with this add-on circuit, of providing a stable output voltage over reasonably wide changes in load resistance. If you can purchase the zeners through industrial surplus channels, you will probably find that this entire project costs less than $5.00, including the mounting case.

PROJECT 13
TRANSCEIVER POWER SUPPLY

In the sixties and seventies, the radio transceiver became the accepted norm among amateur radio operators. These devices contained the transmitter and receiver in the same compartment and shared different parts of its circuitry with both modes of operation. The majority of these used some solid-state circuitry, but they also incorporated tubes of the receive type and often television sweep tubes for the transmitter output amplifiers.

Due to the large variety of components incorporated in these complex transceivers, a number of voltages were required for circuit operation. Generally, a 12-volt source (ac or dc) was used to provide filament voltage. This was also sampled by internal regulator circuits which supplied the proper low voltage to the transistorized stages. These internal circuits also contained their own regulators. Next came the medium voltage for the plate supplies of the receiving tubes. The normal value here was between 300 and

400 volts. A value of about 800 volts was needed for the plates of the television sweep tubes and a bias voltage of between 100 and 150 volts was also required here. Sometimes, the bias voltage was sampled internally from the medium voltage supply. This would depend upon the transceiver manufacturer.

Believe it or not, a single transformer was often used to efficiently supply all of the operating voltages required by these transceivers. During this era, a purchased supply cost about $150. Alternately, you can make one, often for less than $25, by using salvaged components.

Figure 6-30 shows the basic schematic diagram. Notice that a single transformer is used. This is a type which was most popular in black and white television sets of the 1950s and 1960s and can still be found in junked sets today. While there were more junked black and white televisions around in the sixties and early seventies than there are in the eighties, a careful search at your local television repair shop will certainly turn up a number of these which can be had for the asking. Typically, they contain a secondary winding which is center-tapped. The total winding produced between 700 and 750 volts Vac with readings from either side to center tap of 350 to 375 volts ac. Also, there was usually one or two filament windings. These varied from transformer to transformer. One unit might contain a 12.6-volt winding and a 6.3-volt winding, while another might have two 6.3-volt windings. Another type might contain a 5-volt winding for the tube-type rectifiers and a 6.3 or 12.6 volt winding. Some even contained two 12.6-volt windings. It really

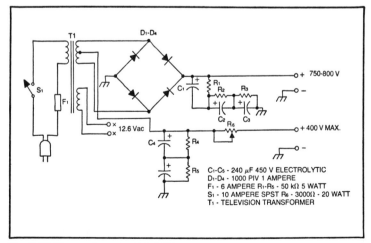

Fig. 6-30. Transceiver power supply.

doesn't matter what values the low-voltage windings are, as it is almost always possible to obtain the needed 12.6 volts. Even when there is just a 6.3-volt winding and one of 5 volts value, it is a simple matter to connect the two windings in series for an output of 11.3 volts, which is close enough. Nearly every one of these transformers contained a 5-volt winding. This is not shown in our schematic drawing simply because it usually has no use in this type of power supply.

The primary circuitry consists of a 10-ampere single-pole single-throw switch and 6-ampere fuse. Television transformers were designed for continuous operation at an output power of about 150-200 watts. Used in intermittent duty for amateur purposes, 500 watts of power could be safely drawn without the transformer becoming overheated. At 500 watts consumption, the primary current will be a little less than 5 amperes.

The rectifier circuit is one that has been used in a previous project. This is known as an economy rectifier or economy power supply, in that it is a full-wave bridge design coupled with a full-wave center tap. The bridge output acts upon the entire secondary voltage value, while the center tap produces an output of half this. The output from the bridge rectifier circuit after it has been filtered is approximately 750 to 800 Vdc under load. This value will rise to 1000 Vdc when the circuit is not loaded. The medium voltage is obtained from the transformer center tap. This will be equivalent to half the high voltage but may be further adjusted with the aid of a 20-watt series resistor which can lower the output to 250 volts or less.

The filter for the high-voltage supply is composed of three 240-microfarad electrolytic capacitors wired in series. Each is rated at 450 Vdc and should contain an insulated case. Obviously, 450 Vdc rating is not adequate to be placed across a 750-800 Vdc source. By combining two units in series (each rated at 450 Vdc), a single capacitor is formed with a rating of 900 Vdc. At the same time, the total capacitance of these two units is halved. Since the voltage output will rise to 1000 Vdc under no-load conditions, yet another capacitor is necessary to give us a total working voltage rating of three times 450 Vdc, or 1350 volts. The total capacitance is figured by dividing the value of one capacitor by the number in the string, or three. This works out to approximately 80 microfarads. Two other capacitors of the same value are used to form the filter network for the medium-voltage supply.

Notice that each of these capacitors is paralleled by a 50,000-ohm, 5-watt resistor. Each of these devices serves two purposes.

First of all, each charged capacitor will discharge into its resistor when the supply is activated. This removes potentially lethal energy from the supply during deactivation. Secondly, each resistor equalizes the internal resistance of its capacitor. This means that each capacitor will drop the same voltage value and no single one will be subjected to a value which is outside of its maximum ratings. Under no circumstances should these resistors be omitted.

The 12.6-Vac supply comes directly from this winding on T1. Since most transceivers do not require dc voltage for the filaments (either ac or dc would work), there is no sense in rectifying and filtering here.

This still leaves us with the bias power supply, which has to be negative with respect to circuit ground. It must also have the ability to be varied from a low value to about 150 volts dc. Figure 6-31 shows how this was obtained. The 12.6 Vac winding was used as the primary drive for a separate filament transformer. Normally, this transformer was designed to be powered from 115 Vac and produce a 12.6 Vac output. However, here we use the secondary winding as the primary and it is driven from the 12.6-Vac output from T1 to produce an output at its secondary of 115 Vac. Often, if a separate 5-volt and 6.3-volt winding was available on the power transformer (in addition to the 12.6 Vac winding), the two would be wired in series-aiding and used to drive T2.

The output from the T2 secondary is rectified and filtered by C5. R7 is a 100 kohm 5-watt variable resistor which allows the bias voltage to be accurately set. This value and the value of the medium-voltage supply are adjusted when the power supply is under load.

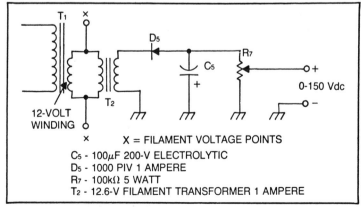

Fig. 6-31. Extra transformer (reverse connected) provides negative bias supply.

Amateur radio transceivers are not the only devices which require operating voltage values that can be produced by this type of supply. When completed, this project makes an excellent addition to any bench and can be used to power many different types of vacuum tube circuits, including radio frequency amplifiers, radio receivers, PA systems, audio amplifiers, and many others. The most expensive items in this supply should be the filter capacitors, assuming that you can pick up the power transformers from a junked black and white television receiver. Some color receivers may also produce a transformer with secondary outputs which are usable for this project. T2 may be the tiniest filament transformer you can find, since very little current will be drawn from this circuit. You should be able to pick this up new from your hobby store for a few dollars. The filter capacitors may be obtained through some surplus channels. The capacitance values are not critical, but use components rated for at least 100 microfarads. Remember, the total capacitance presented by the filter circuit is equal to the capacitance value of one unit divided by three. I have seen computer-grade electrolytic capacitors which would be suitable for this power supply in war surplus catalogs for $1.50 each. Of course, prices will vary from dealer to dealer and will be dictated by supply and demand. If purchased new, the specified units will cost nearly $10 each. Obviously, this will quickly remove any economic advantage of building your own supply. By looking through surplus channels, however, you are bound to come up with a good buy.

Construction of this power supply is quite simple and involves mounting the power transformer on a heavy-duty chassis and installing the primary circuitry. If discrete rectifiers are used, they may be installed on a piece of perforated circuit board. Alternately, an IC rectifier package may be used. Make certain the electrolytic capacitors are insulated. Do not allow them to come in contact with each other nor with the power supply chassis, which serves as ground. This is especially true of the top two capacitors (C1 and C2) in the high-voltage string and C4 in the medium-voltage supply. Even the insulated capacitors are only insulated to a value just higher than their working voltage of 450 Vdc. In a series string of capacitors, each drops a portion of the total output supply voltage. While C1 is dropping only 250 Vdc (well within its ratings), its case will be at a 500 volt or more potential with circuit ground. If you use computer-grade electrolytics, mount them to a large piece of perforated circuit board which will support them away from the power supply chassis. Alternately, you may epoxy circuit board sections to the bottom of each capacitor and then, in turn, attach them to the chassis so that they rest on their circuit board insulators. The bottom capacitor in each string need not be insulated from the chassis but should not come in contact with the capacitor immediately preceding it in the string.

Computer-grade electrolytic capacitors often contain screw-on terminals. If this is the case, the bleeder resistors may be attached at this point, along with the rest of the circuit wiring. Drill holes in the power supply cover in order to mount R6 and R7. Another hole is required for mounting the single-pole single-throw switch. You may also want to install a 115-volt panel lamp to indicate when the power is on.

This supply goes together just like all of the others, except that larger components are being used. Also, a short-circuit or a wiring error will often tend to cause a few more problems, due to the higher voltages delivered by this supply. Other than these differences, it can be seen that this transceiver power supply is just a simple economy circuit with a capacitive input filter and no electronic regulation. Regulation at these high voltages is extremely difficult and is fortunately unnecessary for most applications. The filter capacitors themselves provide enough dynamic regulation for proper operation of most devices driven by this power supply circuit.

Before checkout is attempted, be absolutely certain of your wiring. Take special precautions to check for polarity reversals, open circuits, short circuits, faulty components, etc. Once you are sure that your supply is not plagued by any of these faults, connect a dc voltmeter capable of reading 1000 Vdc or more across the high-voltage output. Activate S1 and note the reading. If the lights dim and the fuse blows, this is a sure sign of a short circuit or wiring error, but chances are that this simple supply will work correctly the first time. The output should be somewhere around 1000 Vdc, although this can vary by 100 volts or so, depending upon ac line voltage and the exact value of the transformer secondary. Carefully remove the voltmeter probes from this point and place them across the medium-voltage output terminals. Your reading here should be half of the high-voltage reading. Series resistor R6 will have almost no effect at this point because the medium-voltage circuit is not under load. Next, check the output of the bias circuit. Adjusting R7 should have a dramatic effect on the output voltage here.

That's all there is to it. If you experience problems with the first circuit reading, most likely the medium voltage will be out of order as well. There is really very little that can go wrong with this type of circuit, assuming that the rectifiers are not defective and have been wired properly along with the capacitors. Of course, polarity reversals will cause the power supply to become inoperable and may damage one or more rectifiers and/or capacitors.

The reader is cautioned to beware of a defective power supply with this voltage potential. Even if you get no high voltage output reading on the first try, assume that the capacitors are fully charged and potentially deadly. Before getting into the power supply, place the shaft of a screwdriver across each capacitor terminal to discharge it. Do this while you grasp the insulated handle. Chances are, there will be no charge in the capacitors, but the flash of an electrical arc is a sure indication that energy is present. Do this every time you work on your power supply. Many persons have been killed by ignoring this simple rule.

Once your power supply has been fully checked and is operational, you can connect it to your transceiver and adjust R6 and R7 (with the aid of an external dc voltmeter) to the proper values. You will probably find that this supply is as good as any that you can buy commercially and is far less expensive.

PROJECT 14
DRY CELL REPLACEMENT

There are still many electronic games and toys and a few calculators and other devices that operate from a single dry cell battery. Additionally, many devices use a number of AA cells of the rechargeable type. It is often difficult to build a power supply with an output voltage equal to that of most single cell batteries, but by using IC regulators, the circuit is no more complicated than a single unregulated supply. The circuit shown in Fig. 6-32 will produce an output of 1.25 Vdc. This is about equal to the voltage of most rechargeable AA batteries. Most dry cells have a 1.5 volt output, but the author has yet to see a device which is adversely affected by operating from a value that is a quarter of a volt lower, as is the case with the output of this supply.

The heart of the circuit is an LM117, three-terminal positive voltage regulator. It is capable of supplying about 1½ amperes of current over a range of from 1.2 to 37 volts. The LM117 is packaged like a standard transistor and is easily mounted to a heatsink for maximum current drain. This supply was originally built to provide a current drain of a few hundred milliamperes, so it wasn't neces-

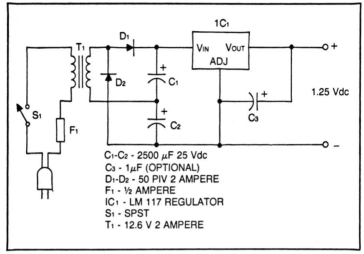

Fig. 6-32. Low-voltage power supply circuit.

sary to mount the IC to a heatsink. However, if you desire the full 1.5 ampere output, the heatsink will be mandatory, and it will be necessary to increase the transformer power rating to about 4 amperes. Five-ampere diodes should be used for maximum current output.

A 12.6-volt transformer is used in a voltage doubler circuit to produce an output of about 30 volts. Note the standard full-wave voltage doubler configuration consisting of D1 and C1 with D2 and C2. The output from this circuit is fed directly to IC1, the LM117. This IC is packaged in several different ways. Sometimes, pin 1 is the input, while pin 2 is the ground and pin 3 is the output. Another configuration allows pin 2 to serve as the input, pin 1 the ground and the case as the output. The IC is labeled in the schematic diagram as V_{in}, V_{out} and ADJ. The data sheet you get with your purchased IC will identify the pins which go with these designations. The IC draws power directly from the power supply unregulated output and produces a regulated 1.25 Vdc. C3 may be omitted. It is designed as a filter to remove transient voltages. It would be possible to up the output voltage of this supply to an even 1.5 Vdc, but this would require the addition of two resistors and is really not worth it. Another project in this text describes a variable voltage supply using this same IC which can be used if you desire the slightly higher voltage. As is, this power supply can be used to power almost any device which is presently being driven by a AA cell. Additionally, it can be used to charge these small batteries.

Begin construction by mounting T1 and its associated primary circuitry in a small aluminum box. D1 and D2 in the secondary may be mounted on a small piece of perforated circuit board. Alternately, a larger circuit board may be incorporated to mount the two rectifiers and the two filter capacitors. The latter are large units owing to their high capacitance, and you may prefer to mount them against one wall of the power supply chassis or case. If the IC is to be used with a heatsink, you may use the aluminum case for this purpose. Most of these ICs are sold with the appropriate mounting hardware, which includes the insulator. In many configurations, the device case will serve as the positive output terminal and must be isolated above ground. When this is done, be sure to use liberal amounts of silicone grease smeared over the mounting surfaces to assure good thermal conductivity. The supply is completed by installing the output terminals.

Look over your finished circuitry carefully. Make sure C1 and C2 are wired in series with the negative lead of C1 connected to the positive lead of C2. Make certain D1 and D2 are not reversed and be sure your IC is wired according to this schematic. Mixing the IC leads can completely destroy the device.

Connect a voltmeter across the output terminals. Sometimes, it is difficult to obtain an inexpensive meter which will accurately measure voltages in this range. For a true test of operation, it may be necessary to borrow a voltmeter with a 1.5-2 volt scale. When the power supply is activated, the meter should indicate 1.25 Vdc. This value may be fractionally higher or lower, as no two devices are exactly alike.

That's all there is to it. This is such a simple circuit that it should work the first time and every time. If you do experience problems, temporarily disconnect the lead between the positive side of C1 and the Vin IC terminal. A dc voltmeter placed with the positive probe here and the negative probe at ground should indicate approximately 30 Vdc. If you get nothing at this point or if the voltage is very low (20 volts or less), you have a wiring error or a defective component in the doubler circuit. Check the secondary voltage output of the transformer with the voltmeter in the ac position. You should get a steady 12.6 Vac here.

On the other hand, if the output from the voltage doubler reads the nominal 30 Vdc and you definitely have no wiring errors associated with the integrated circuit, then the IC has been damaged and must be replaced with a new unit.

Again, maximum current drain from the supply as shown should be limited to a few hundred milliamperes. By mounting the

IC to a heatsink and increasing the current ratings of T1 and the rectifiers, a maximum of 1.5 amperes may safely be drawn. When operating at maximum current output, monitor the temperature of the integrated circuit case. If it seems to be getting too hot, a larger heatsink is called for. The manufacturers' ratings will often contain the physical size of the heatsink. This is printed along with other device data on information sheets that are usually included with the device upon purchase.

Once the supply is operational, you can connect it to any device which does not draw more than the maximum current this circuit is designed to deliver and which was previously powered by a dry cell battery. It will not be necessary to constantly replace discharged batteries while operating near a source of alternating current. For charging batteries, simply connect them in parallel with the output of the supply (positive to positive/negative to negative). Make certain that only rechargeable batteries are connected to this supply, as other types may overheat or even explode, creating a potential fire hazard. If you've built the supply for maximum current output, you should be able to charge six or more batteries simultaneously. To make this supply really fancy, you might wish to install an in-line ammeter. This would be installed in series with the positive output leads and should be capable of reading 0 to 2 amperes of dc current. This is especially helpful when charging batteries, as it allows you to know exactly when the batteries have reached their full potential.

PROJECT 15
5-VDC 1-AMPERE IC SUPPLY

Some types of integrated circuits require a highly stable 5 Vdc input. Here, it is necessary to use highly complex regulator circuits to hold the voltage swing to a very minimal value. Such supplies used to be very expensive to construct due to the large number of parts that were required. However, in this day and age, many of the discrete components have been completely replaced with a single integrated circuit which derives all of its operating power from the unregulated power supply and produces a regulated output.

Such is the case with the power supply shown in Fig. 6-33. The heart of this circuit is an LM309K regulator, which has only three contact points. Pin 1 is the voltage input, pin 2 is the regulated output terminal and pin 3 connects to circuit ground. It is necessary to drive this IC regulator with as stable a voltage as possible.

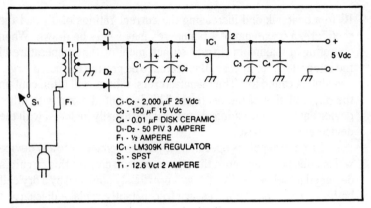

Fig. 6-33. Five-volt, one-ampere supply circuit.

Therefore, two filter capacitors are used in parallel at the output of the full-wave center-tapped rectifier. Capacitors C1 and C2 are 2000-microfarad units rated at 25 Vdc each. When connected in parallel as shown in the schematic drawing, they form a single capacitor rated at 4000 microfarads. The voltage rating is still 25 Vdc. I have found that it is often easier to obtain two capacitors of this type than a single capacitor rated at 4000 or 5000 microfarads. Naturally, if you find it just as convenient to use a single large capacitor, do so. If you have access to capacitors with even higher values, these are fine and will even provide a little better stability, although the capacitors specified in the schematic drawing are certainly more than adequate.

The output from the filters is connected directly to pin 1 of the integrated circuit. Pin 3 is connected to the power supply ground, and pin 2 is connected to the output circuitry. This consists of a 150-microfarad electrolytic and a 0.01 microfarad disk ceramic capacitor for the suppression of transient voltages. These latter two capacitors may be omitted from the circuit but are placed there to provide the best overall performance possible to be suited for the most critical of applications.

Construct the power supply as you would any unregulated type. It will probably be best to mount the filter capacitors to a large section of circuit board which can also serve as a base for the diodes and the two remaining capacitors. The integrated circuit must be mounted to a heatsink which can be the power supply case. Make certain that the IC is mounted in close proximity to C1 and C2. This circuit is capable of producing about 750 milliamperes of current at 5 Vdc. However, if you intend to draw only 100 milliamperes or so, the heatsink mounting can be avoided and all components can be mounted to a single section of circuit board. You can also decrease the current rating of the power transformer and the diode by a factor of 50%.

This is a highly complex power supply which has been made extremely simple by the fact that the majority of complex circuitry is housed entirely within the IC case. All the builder has to contend with are the three connections to a rather basic power supply circuit.

There are no adjustments to be made to this circuit. Once you have completed the assembly, examine it carefully to make certain you have not inadvertently reversed any leads or connections, especially around the integrated circuit. When this supply is activated, it either works or it doesn't. Connect a dc voltmeter at the output terminals and activate S1. You should get a reading of exactly 5 Vdc. If not, it will be necessary to remove the connection between C2 and the integrated circuit. Take a voltage reading across the terminals of C2 with a voltmeter which is capable of reading 15 volts or more. An appropriate reading at this point will be approximately 10 Vdc. If you get no reading at this point, measure the secondary ac voltage across C1. This should be 12.6 Vac. Now, check the readings between the center tap and either side of the transformer secondary. 6.3 volts should be read on both sides. If the transformer appears to be functioning properly, then your problem probably lies within the rectifier circuitry. Alternately, one or more filter capacitors may be defective.

If the output from the filters appears normal (about 10 Vdc), disconnect C3 and C4 from the supply circuitry. If you get a near normal reading then, one of these latter components is defective. If you still get no reading, then the integrated circuit regulator is not operating and should be replaced with a new unit.

I have found that it is a very rare occurrence indeed when an IC regulator does not operate fresh from the box. Nine times out of ten, IC inoperation is caused by improper connection to an activated circuit. Integrated circuits can be damaged in far less time than it takes to blow a line fuse. For this reason, it is mandatory that you examine all wiring before initially activating a newly built power supply. IC voltage regulators can sometimes be the most expensive components in any power supply circuitry and are quite costly to replace. In nonregulated supplies, the power transformer is often the most expensive component, but these devices are very rugged (when compared to IC regulators) and are not usually destroyed in a fraction of a second.

Once the power supply is operational, it may be connected to any electronic circuit which demands extremely stiff regulation and high ripple rejection. A supply of this sort will often provide regulation figures of 0.1% per volt or better. This is usually adequate for even the most demanding of electronic loads.

PROJECT 16
SERIES-REGULATED DUAL-POLARITY SUPPLY

Many dual-polarity power supplies designed for the operation of linear integrated circuits and operational amplifiers use simple zener diode regulator circuits to stabilize output voltage. Often, this is quite adequate, but in some instances, the degree of regulation is not sufficient to provide suitable operation. Here, it is necessary to resort to the more complex series regulator circuit which has been used in several other projects in this book. Series regulation provides a high degree of stability and will satisfy most operating requirements.

Dual-polarity power supplies provide a positive potential at one output terminal with respect to circuit ground and a negative potential to ground at another terminal. Since there are two distinct voltage legs, it is mandatory that we provide separate regulator circuits at each. Figure 6-34 shows a power supply which will deliver a 15-volt dc potential at both positive and negative polarities with respect to circuit ground. To do this, two full-wave center-tapped diode circuits are attached to the output of a single power transformer. The operation of this circuit was explained earlier, but to recap, this is not a full-wave bridge circuit, although the two appear similar. The center tap is used here and no ground is provided at the back diodes of the apparent bridge. Again, this is two full-wave center-tapped circuits (one set up for positive conduction and the other for negative) connected to the output of a single 25.2 Vct transformer.

Rectifiers A and B pass those portions of the ac cycle which are positive in nature while rejecting negative potentials. The pulsating positive dc output is filtered by C1, a 1000-microfarad electrolytic capacitor. The smoothed output is then series-connected to the regulator transistor, Q1, which conducts just enough to pass along a 15 Vdc output. R1 sets up the proper operating parameters for the base connector circuit of Q1. The zener diode serves as a reference device and is also part of the base circuit, along with C3. The

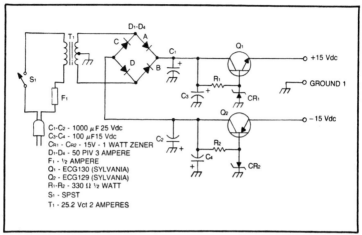

Fig. 6-34. Series-regulated dual-polarity supply.

resultant output is an extremely stable 15 Vdc in relation to circuit ground.

Looking at the other half of the circuit, rectifiers C and D pass the negative portions of the ac sine wave and conduct pulsating negative dc to filter C2. This smoothed output is passed along to Q2, which is a PNP transistor (the reverse of Q1). R2, C4 and CR2 establish the base circuit of Q2, just as the equivalent components did in Q1. The resultant output is negative 15 Vdc in relation to circuit ground.

This circuit should be installed in a small aluminum box. Mount the transformer and primary circuitry first. Then prepare to wire the secondary on a single section of perforated circuit board. A 4" × 5" section should be adequate, but you can shrink this quite a bit by closely grouping the components. Figure 6-35 shows the arrangement I used. However, wiring is certainly non-critical and you may wish to use a different component arrangement. There are many polarized devices in this circuit, and since the positive portion is a mirror image of the negative circuit, it is quite easy to inadvertently reverse a component if you're not extremely careful. It is best to begin by wiring the four diodes in place. Using an IC bridge package will simplify this procedure further. Identify the positive output connection from the rectifier circuit and wire the remainder of the positive leg. This includes the filter, the transistor and its associated circuitry. Upon completion of this stage, reexamine all components and connections to make certain that no reversals have

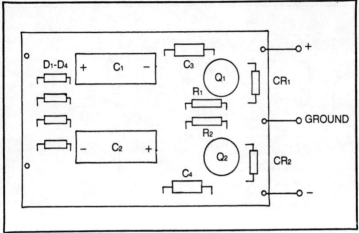

Fig. 6-35. Component layout on perforated circuit board.

been allowed to occur. Note that the positive terminals of C1 and C3 are connected directly to the rectifier output line. Note that the cathode of CR1 connects to the base electrode of Q1 and to one end of R1.

Once you are satisfied with the positive leg of the circuit, you may begin on the negative side. Identify the negative takeoff from the rectifier circuit (the junction of rectifiers C and D). From this point to the negative output, wire the complete output circuitry. It is sometimes difficult for beginning builders to realize that the positive terminal of a capacitor is often grounded. True, it is more often seen with the negative terminal serving as ground, but the reverse is certainly encountered from time to time. This is one of those situations. Remember, you are working on a mirror image circuit of the positive leg you have just completed. Here, the positive terminals of C2 and C4, along with the cathode of CR2, are connected to circuit ground. The reverse was true of the previous leg.

Once you have completed wiring the negative circuit, reexamine it closely to make certain you have not transposed any connections. Then, compare it with the positive circuit leg, making sure that all polarized components are reversed (mirror image). Your circuit is now completed.

Mount the circuit board near the power transformer and make the required connections to the input. The output from the circuit board should be connected to the appropriate output terminals, which will most likely be mounted through the chassis wall. The aluminum chassis itself may serve as a ground connection, or you may wish to provide a separate ground conductor, allowing circuit ground to float above the chassis.

For the checkout procedure, connect the positive probe of a dc voltmeter across the positive output terminal. The negative probe is attached to circuit ground, *not* to the negative 15 volt leg. Activating S1 should produce a reading of 15 Vdc. If not, quickly deactivate the supply and attach the positive probe to circuit ground and the negative probe to the negative 15 Vdc output terminal. Reactivate the supply again and note the reading. If you get nothing, then the power supply has a problem which both circuit legs mutually share. If, however, you get a reading at one leg and none at the other, a problem which pertains only to that portion of the circuitry used by the dead leg exists.

If you have no voltage at either terminal, suspect damaged or reversed rectifiers, a defective transformer, a blown fuse or a lack of line voltage. If the positive leg is operational while the negative leg is not, suspect rectifiers C and/or D, a defect in C2, C4, R2, Q2 or CR2. Should there be proper voltage at the negative leg and none at the positive leg, suspect rectifiers A and/or B, and the remainder of the components in the positive leg. Remember, a reversal of polarity of any of these components may cause destruction of the reversed components upon initial activation. It may be necessary to replace all improperly wired components before attempting the checkout procedure again.

When the power supply is operational, you will possess a circuit which provides very stable dc output voltages of positive and negative polarities with respect to ground. Using the components specified, it should be safe to draw approximately 600 milliamperes from each circuit leg. Up to 1 ampere may be drawn from either leg, as long as the drain from the other is 200 milliamperes or less. Remember, this dual-polarity supply shares a common transformer, so output current from both legs adds to the total drain from the supply. You will find this power supply to be adequate for driving a wider range of more critical electronic devices than is possible with the simple zener diode-regulated, dual-polarity supply discussed earlier in this chapter. Both transistors, Q1 and Q2, are designed to be mounted to a heatsink. However, they are overrated for this application. If you stick to the output current specified here, they will operate well within their thermal limits and will provide many long years of service.

PROJECT 17
300 VDC WITHOUT A TRANSFORMER

Did you ever stop to think how small and lightweight the larger types of electronic equipment might be if you could avoid using a power transformer? In large audio and radio frequency amplifiers, a transformer may weigh thirty pounds or more and encompass a large area of space within the device chassis. In some applications, it is possible to do away entirely with the power transformer and still provide moderate to high levels of dc output voltage. How do we do this? By using voltage multiplication *instead* of voltage transformation. To transform voltage, we must start with a source of alternating current. Voltage multiplication also involves ac, and there is a ready supply at every wall outlet. Often, however, the 115 Vac (or 230 Vac) is far below the potential needed to operate certain types of electronic equipment. Here's where voltage multiplication comes to the front and provides proper operating values in a minimum of space and with great weight reduction when compared to transformer-type circuits. Circuits which utilize this type of construction are known as line-derived power supplies and a full-wave voltage doubler type is shown in Fig. 6-36. The output from this power supply is approximately 300 Vdc under load. However, if you draw a maximum current of 2.5 amperes, the value will probably drop to closer to 250-275 Vdc. This amount of current, however, is equivalent to nearly 700 watts and yet the total circuit should weigh in at only a few ounces.

Examining the schematic closely, we see that it is the old standard full-wave voltage doubler circuit that has been used several times before in this chapter. The main difference here is that the power transformer is not used and the ac input is derived directly from the wall outlet. You can think of the wall outlet as being the secondary of a power transformer, which delivers 115 Vac RMS. This is indeed exactly what it is, but the power transformer is mounted on a pole near your home.

The largest components in this circuit will be C1 and C2. I used 250-microfarad units rated at 250 Vdc, although to conserve space, 80-microfarad units would suffice. For most applications, it is best to use as high a value of capacitance as possible to promote good dynamic regulation. The 80-microfarad capacitors, however, could probably be mounted on a medium-sized section of circuit board, whereas the 250-microfarad units will most likely require mounting on or under a chassis. This adds to the circuit size. In either case, D1

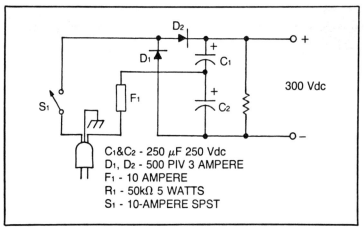

Fig. 6-36. Transformerless power supply.

and D2 would be mounted to a small section of circuit board, but could be mounted directly to a small terminal strip and the terminals of C1 and C2. R1, the 50 kohm 5-watt bleeder resistor, is mandatory for use with this circuit and will discharge the capacitors when the circuit is deactivated. F1 is a 10-ampere line fuse, while S1 is a single-pole single-throw toggle switch rated to handle the 10-ampere maximum. Both of these components are actually over-rated, as it is not safe to draw more than about 2½ amperes of current from the output of this supply. The limitation here is the current rating of the two rectifiers, which was limited to 3 amperes. These could easily be replaced with 5-ampere units or even 10-ampere types for greatly increased power output capability. Remember, the 115-volt ac line is derived from the secondary of an extremely large power transformer owned by your electric company. In modern homes, most of these outlets are protected by a 30-ampere fuse or circuit breaker, which means that as much as 3000 plus watts may be drawn from this source without exceeding design limitations. However, regulation becomes a severe problem when current drains approach these maximums. While live-driven multiplier circuits may be used for high power applications, this is rarely done except in cases where a steady current drain is present rather than a constantly fluctuating one.

The line fuse is absolutely mandatory. This circuit has the potential of drawing a lot of current should a short circuit occur. While the current is high, it might not be adequate to trip the circuit breaker which protects the receptacle output, and a fire could begin within this power supply.

Treat this circuit as you would any of the voltage doubler circuits previously discussed. The leads from the line cord will be treated in exactly the same manner as the secondary leads from a power transformer and are connected to the junction of D1 and D2 and C1 and C2. The bleeder resistor specified here is designed for a 300 Vdc output and will discharge the circuit when it is deactivated. It also presents a minimum load to the power supply at all times and prevents large swings in output voltage.

The circuit should be mounted in a metal enclosure which is attached to the household ground system. This is accomplished by using a three-pronged, polarized plug, the ground connection of which is attached to the chassis. You may also elect to use shielded cable between the plug and the circuit to prevent stray noise pickup. This will help in some situations, but a line-derived power supply of the transformerless variety does not offer the line isolation that can be obtained by the insertion of a transformer. For this reason, audio equipment driven by such a power supply will often pick up a fair amount of line noise. Much of this can be filtered out, but filter components make the circuit far more complicated.

Once you have completed your assembly, examine it closely for signs of short circuits. Be sure to perform a thorough inspection, because a short here will often cause arcing and component burning. Insert the polarized plug in the wall outlet and with a voltmeter across the output, activate S1. Depending upon your line voltage, you should get an output reading somewhere around 300 volts. If your line voltage is exactly 115 Vac, then the output will read closer to 320 volts. If the line voltage is higher, the output will be higher as well. If lower, this would be reflected at the dc output.

This is one of the simplest circuits presented thus far, since it uses a minimum of components. If you fail to obtain a 300 Vdc output reading, this is probably due to a defective diode, a blown fuse, electrical failure or a broken conductor. Also, D1 or D2 might be reverse-connected in the circuit. On the other hand, if the fuse blows, this is an indication of a short circuit which might be caused by a defective component or a wiring error. Such a short may also cause D1 and/or D2 to be destroyed, so if failure occurs due to a short circuit, it will be necessary to check each component to see that it is still operational.

That's all there is to it. Just remember that the output from this power supply is potentially lethal, as is the input from the 115 Vac line. If it becomes necessary to troubleshoot this power supply with the power on, be extremely careful. Remember, no point in this circuit can be thought of as low in voltage. Every part of this circuit, from the line plug to the dc output, has the capability of causing physical injury and death. The purpose here is not to be morbid but to encourage the reader who may be more familiar with low voltage power supplies to think safety first and constantly.

PROJECT 18
600 VDC WITHOUT A TRANSFORMER

This project is very similar to the previous one, except the output voltage is doubled. The circuit looks identical. Only the component ratings have been changed. Figure 6-37 provides the new components list, and you can still use the previous schematic. Just make the substitutions listed here and your supply will be ready to go.

The reason for the increased output is the fact that instead of being attached to the 115-volt ac line, this circuit is designed to receive driving power from the 230-volt line. Of course, you will have to replace the line plug in the previous schematic with that designed for a 240-volt receptacle. Such receptacles will be found near electric clothes dryers, electric ranges, etc. These tend to be rather large, so you may wish to use one of the many different types of plug and receptacle combinations to supply power for this circuit. Never connect a standard 115-volt line receptacle to a 230-volt source. This, in itself, is not unsafe, but the fact that it looks like any other 115-volt receptacle is. Should someone attempt to plug in a 115-volt device to this receptacle, which is supplying twice this value, the device might be damaged irreparably and a fire could even result. Most 230-volt plugs and receptacles are designed so that they cannot be intermixed with 115-volt types. The two types are not compatible.

Even though this circuit doesn't use a transformer (at least not within the power supply proper), we can think of this supply as being driven by a secondary output voltage which is twice that of the previous circuit. This is the reason for the increased voltage output.

C_1 & C_2 - 250 µF 450 Vdc
D_1 & D_2 - 1000 PIV 1.5 AMPERE
F_1 & F_2 - 10 AMPERE
R_1 - 10 KΩ 5 WATT
S_1 - 10 AMPERE SPST

Fig. 6-37. Components list for transformerless supply.

In looking at the new components table, we see that the capacitance value of C1 and C2 has remained the same, but the working voltage rating has been doubled. The peak inverse voltage rating of D1 and D2 has also doubled, while the current rating has been halved. F1 and S1 remain the same, while the ohmic value of R1 has been doubled. The power rating of the bleeder resistor remains the same.

This circuit will deliver the same amount of power as the previous one if you use the components specified. This is somewhere within the range of 600 watts. Safe current drain with the previous supply was 2.5 amperes. With this supply, you should limit current drain to about 1 ampere. In other words, this supply has doubled the output voltage, but approximately halved the current. Since power is a product of voltage and current, doubling one factor and halving the other results in the same output.

This power supply was designed to operate from the 230-volt mains. The neutral conductor is not used. Notice that an additional component, F2, has been added. This is another 10-ampere line fuse, which is inserted in the remaining 230-volt leg. In other words, both conductors from the 230-volt source will be fused. This provides additional safety in case of a short circuit and prevents a possible shock hazard.

Circuit completion and checkout is handled exactly as in the previous project. As a matter of fact, except for the line plug arrangement and the slightly larger physical size of the filter capacitors, this circuit should look identical to the previous one.

The output should read approximately 600 to 650 volts under no-load conditions and can be used to provide operating power to the plate circuits of vacuum tube devices and for other such applications. As before, you may increase the current ratings of any of the series components for increased power output, but be careful not to exceed the design limitations of your power system.

This power supply offers a bit of a bonus over the previous one, in that it will also deliver 300 Vdc output if you simply change the line plug to a 115-volt type and connect it to this lower voltage source. All of the components for this latter supply are overrated for a 115-volt input. Therefore, you can safely switch to the lower voltage without any adverse effects. Remember, however, to observe the approximately 1-ampere limit on current drain, since the rectifiers are rated for a maximum of 1.5 amperes. Using the components specified, the supply will deliver only half the power output of the previous project when operated in the low voltage mode. The bleeder resistor will take a little longer to completely discharge the capacitors, since it was originally set up for the 600-volt value. This added time, however, is measured in seconds and should not create an undue safety hazard.

PROJECT 19
IC-CONTROLLED VARIABLE VOLTAGE SUPPLY

Most of the power supplies discussed thus far in this chapter have provided an output voltage of a single, fixed value. A few have provided dual voltage outputs, and others provided dual voltage and dual polarity capabilities. These are adequate for many applications, but are also inconvenient when it is necessary to power a number of different devices, many of which require different driving values. For example, it is not unusual to encounter electronic circuits which may require 3.9, 5, 6, 7.5, 9, 12, 15, 18, etc. volts. While it is true that most devices require 6, 9 or 12 Vdc, there are many exceptions to this rule, especially when dealing with test circuits which use IC devices. Here, it is necessary to have a larger number of power supplies to be fully capable of a complete design and test operation. Obviously, the expense involved in building eight or ten different power supplies is not practical, nor is making room for this many testing units.

Fortunately, there is an alternative. It's known as a variable voltage supply. This is a fairly modern introduction to electronics test benches when speaking of electronically regulated types. The older type of variable voltage supply has been around for many decades and simply consisted of a basic power supply whose output was controlled by a series resistor and often a resistive voltage-splitting network. It was necessary to make adjustments to the output voltage by varying the output resistance while the power supply was delivering current to the load. Different current drains dictated different resistance settings to deliver the same amount of voltage. This was bothersome in itself and, of course, the output was only regulated by the size of the filter capacitor and filter choke and not by any electronic means. This older type of variable voltage supply would be totally unsuitable for many types of solid-state circuits.

The first electronically regulated variable voltage supplies were extremely expensive, as they were terribly complex. In layman's terms, they incorporated sampling circuits which exercised low voltage control on a series pass transistor network. They operated in basically the same manner as a simple electronic series regulator. The complexities came in when it was necessary to constantly change the output transistor's reference voltage, which determined its amount of conduction. Obviously, conduction had to be lowered for a low voltage output and raised when the output was to be swung higher.

Today, variable voltage power supplies with good electronic regulation are simple projects and certainly inexpensive when compared with earlier models. Integrated circuit technology is the sole cause of these excellent features and typical circuits are no more complex than single voltage supplies. Figure 6-38 demonstrates this fact. This simplistic-appearing circuit is capable of providing a continuously variable output from 1.5 to 25 Vdc with an output current of around 1 ampere. The output voltage is highly regulated and can be used to power some of the most demanding electronic circuits.

You should be able to readily identify the full-wave bridge rectifier circuit, which is connected to the secondary of T1, a 25.2-volt transformer rated for an output of at least 3 amperes. Many of these transformers are available as center-tapped models. If yours has a center tap, simply snip it off or better yet, tape it out of the way, as it may be useful for some later project.

After the secondary ac is rectified, it is filtered by C1, which is rated at 2500 microfarads. This is a relatively large amount of capacitance and serves to keep the voltage input to the integrated circuit voltage regulator very stable. You may get away with using slightly lower values, although this will affect regulation to some degree. Higher value filter capacitors will not give appreciably better performance but may certainly be used without incurring any disadvantage whatsoever.

The heart of this circuit is the LM117 voltage regulator. This was used in a previous project to provide a single output voltage of 1.25 Vdc. In this application, two resistors have been added to the circuit to provide for the variable voltage output. With the components shown, the low value will still be around 1.25 volts, although most persons will elect to use 1.5 Vdc, which exactly matches the output of a dry cell battery.

The output from C1 is connected to the Vin terminal of the IC. The ADJ lead is connected to R2, R1 and Vout is connected to the other end of R1 and, of course, to the positive output lead from the supply.

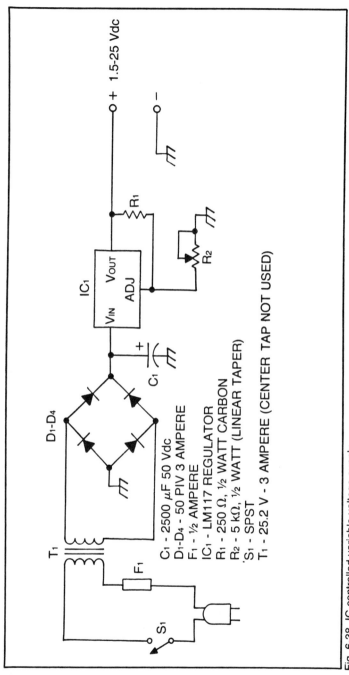

Fig. 6-38. IC-controlled variable voltage supply.

All components may be mounted to a section of perforated circuit board, except for R2 and the integrated circuit. It will probably be necessary to mount the IC on a heatsink if you anticipate drawing the full 1 ampere of current. If you intend to draw only 100 milliamperes or so, the IC can be operated without a sink while staying within its thermal maximums.

R2 should be installed in the aluminum case which will also house the remainder of the circuitry. Drill a hole through the front panel which will accept the shaft of this potentiometer, and then connect to the rest of the circuit using short lengths of stranded conductor. This potentiometer is wired as a variable resistor, which involves shorting out two of its three leads, as shown in Fig. 6-39.

Construction is straightforward, and the placement of components is noncritical, except that the integrated circuit voltage regulator should be placed in close proximity to the filter capacitor. Be certain to observe polarity during every construction step, as a reversal could cause damage to the regulator. Be sure of the three regulator contacts to avoid a reversal here. The output voltage is a function of the values of R1 and R2. Since R1 is a fixed unit, do not substitute other values.

Once your project is completed, examine all solder contacts and re-inspect the circuit for signs of broken conductors, shorts, wiring errors, etc. You are now ready to test for proper operation. First of all, connect a dc voltmeter across the output terminals of the supply. Start with the meter in a range which will read up to about 30 volts dc. With the potentiometer in the fully clockwise position, activate S1. The output voltage should read 25 to 28 Vdc. Now, begin to reverse R2. The voltage should begin to decay, and you should be able to take it all the way down to about 1.5 Vdc in the fully counterclockwise position. This may vary slightly from circuit to circuit.

If you get no voltage reading, reinspect your wiring from the line plug to the IC. Take a voltage measurement at C1 with the IC temporarily removed from the circuit. No voltage here indicates a defective transformer, lack of primary voltage, defective diodes or a possible broken connection between C1 and the line plug. If you have voltage here, reconnect IC1 and try again. If you obtain no output, temporarily disconnect R1 and R2 and ground the ADJ lead. If the IC is operational, you should obtain an output voltage of approximately 1.25 Vdc. If you get nothing, then the integrated circuit voltage regulator is defective and must be replaced. This assumes that you are obtaining a reading of approximately 35 volts at C1.

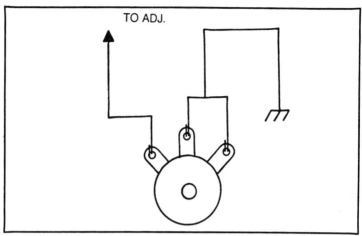

Fig. 6-39. Potentiometer wiring.

Once this circuit is known to be functioning properly, it may be used to power any devices which require from about 1.5 to 25 Vdc with good regulation. However, as shown, it will be necessary to use an external voltmeter to allow you to know what voltage output the supply is producing at the various control settings. It is possible to make a faceplate for mounting around the potentiometer control shaft. When a pointed indicator knob is used with this plate, it is possible to calibrate the two to indicate output voltage (i.e., when the shaft is in the 12:00 position, the dc output is 12.5 volts). This, of course, will not be an extremely accurate method of setting output voltage, and many persons prefer to add an internal voltmeter to the circuit. These can be purchased through surplus channels for a few dollars, although accurate commercial models may cost $30 or more. Some persons also like to add an internal ammeter to allow them to monitor the current drain of any device being powered by this supply. Figure 6-40 shows the correct placement points for the voltmeter and/or ammeter in the previous circuit. Notice that the voltmeter is wired in parallel with the dc output, while the ammeter is in series with the negative output line. Meter polarity must be observed for proper indications.

With the addition of the metering circuitry, this supply becomes even more versatile and valuable for a myriad of electronic test functions. The ammeter is especially useful in diagnosing problems within the equipment being powered and will also indicate when the circuit is being operated near or outside of its maximum current limitations. Such a supply should serve the builder for the remainder of his career.

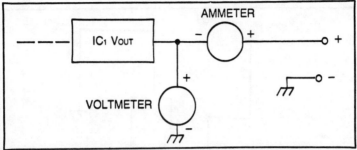

Fig. 6-40. Measurement point for testing power supply circuits with built-in voltmeter and/or ammeter.

PROJECT 20
SOLAR POWER SUPPLY

All of the power supplies discussed thus far have received power from the ac line. This is the most convenient source of readily available power in the United States, with storage batteries coming in at number 2. Another convenient source of electricity is the sun. Certainly, you have heard of a solar cell. This is a device which converts sunlight to electrical power. The solar cell is more accurately known as a photovoltaic cell. Its construction is very complicated and uses a material which generates a current flow when exposed to light waves. The photovoltaic cell is shown in Fig. 6-41 and is made by combining two ultra-thin layers of silicon crystal which has been treated with a specific amount of impurities. While it is not the purpose of this book to discuss solar cell construction in depth, in simple terms, it can be said that one material is negative while the other is positive. When the two are sandwiched together, a PN junction is formed and a photoelectric effect takes place when this junction is subjected to light rays.

Solar cells can be used just like batteries to make a dc power supply. Their output is direct current and they can be combined in parallel or in series to arrive at different voltage and current capabilities. The output current of the solar cell is directly proportional to the amount of light present at the device and the surface area of the cell. When two cells are constructed in the same manner, one with twice the surface area of the other, the larger cell will produce twice the current of the smaller one. This assumes that both are subjected to the same light levels under identical conditions. While output current will vary with the device size, voltage usually does not. Most solar cells produce an output voltage of approximately 0.45 Vdc. While this figure may vary by a tenth of a volt or so due to differences in device construction and materials, this value is the most common. This voltage is the potential difference at the PN junction. To get an increase in voltage, a series connection of several solar cells will be required.

While solar cells can be used just like batteries, it should be understood that they are certainly not as dependable in delivering power. The fault lies not in the solar cell itself, but in the fact that it derives its drive from the sun or an artificial source of light. Even medium power applications will require a great deal of sunlight, and of course, this is impossible at night or on a cloudy day. For this reason, solar cell power supplies are often used in conjunction with rechargeable batteries. The solar electricity is stored within the battery, which then supplies power to the load. In this way, any time the sun is bright enough for the cells to produce electricity, the power is automatically channeled to the battery. This assures a readily available supply of electrical power when it becomes necessary to provide electricity to the load. If you were to drive the load directly from the solar cell supply, you would have operation only when the sun was shining. A temporary shadowing effect due to clouds would temporarily cause the power to fail. Obviously, the potential for intermittent operation here makes the direct use of solar cells as dc power supplies impractical. But by adding a rechargeable battery, the circuit is very useful for providing a continuously available source of free electrical power.

As you learned previously, the output from a solar cell is typically 0.45 volts, regardless of the nominal device current level. This is hardly enough to power even the tiniest transistor radio, so a

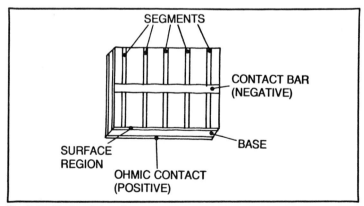

Fig. 6-41. Photovoltaic cell.

means must be found to increase the voltage delivered to the electronic device to be powered. The solar cell closely resembles the operation of a battery, in that it has two poles, one positive and the other negative. As a matter of fact, a solar cell is also called a solar battery, although it does not store energy.

To increase the voltage when dealing with batteries, each component is wired in series with the positive pole of one attached to the negative pole of another, and so on. If we wire three batteries in a series circuit, then the total output voltage will be three times the value of any one, assuming that all three are identical. What about the current value in this series connection? It remains the same. If one battery will deliver 4.5 volts at 1 ampere, then three batteries in a series connection will deliver 4.5 volts at 1 ampere. Only the voltages add in a series circuit. The current rating remains the same.

Let's apply this to solar cells now. A single solar cell exhibits an output voltage of 0.45 volts dc. Figure 6-42 shows how three of them might be wired in series. Notice that the positive pole of one cell is connected to the negative pole of another which, in turn, has its positive pole connected to the negative pole of the third. The three components are wired in series, just as if they were batteries.

For the sake of discussion, we will say that each solar cell is capable of delivering 0.1 ampere to a normal load. Now that three have been connected in series, what will the voltage and current ratings be for this complex circuit? The answer is simple. In series circuits, the voltages add while the current remains the same. So, the output voltage will be equal to three times the value of a single cell, or 3 times 0.45. The output voltage will be 1.35 volts dc. The current rating will be the same as for a single unit, or 0.1 ampere.

By combining three cells in series, we have tripled the output voltage, kept the current capability the same and tripled the power output over that obtainable with a single solar cell. Power is equal to the output voltage times the output current ($P = IE$). A single cell

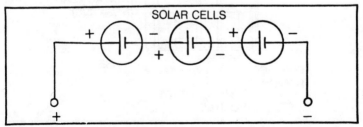

Fig. 6-42. Wiring of three solar cells in series.

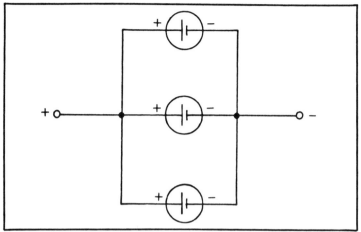

Fig. 6-43. Parallel wiring of three solar cells.

has a power output of 0.45 volt times 0.1 ampere, or 0.045 watts, or about one twenty-fifth of a watt. When three cells are combined, the power output rating is still voltage times current, but the voltage output has been tripled; so the formula reads Power = 1.35 × 0.1, or 0.135 watts, which is equivalent to about one-eighth of a watt.

Series connection is the rule rather than the exception with most solar cells used for power supply purposes. Often, you will find ten or more arranged in a series connection. Even a 6-volt supply would require at least thirteen cells. A 12-volt battery replacement would require that at least twenty-six solar cells be wired in series, and this is the case regardless of the power requirements of the load.

Again, for the sake of this discussion, let's assume that one solar cell delivers 0.45 volts at 0.1 ampere in the bright sunlight. Assuming that the voltage output level presents no problem, how can we go about increasing the current rating? From the earlier example, we already know that three cells are capable of supplying about an eighth of a watt of power. Could we wire them in such a way as to have the output deliver the same 0.45 volts but at 0.3 ampere, three times the current of a single cell? Certainly. When solar cells are connected in series, the voltage of each unit adds to the total circuit voltage, but the current remains the same. However, when you wire the cells in parallel, the current adds while the voltage remains the same.

Figure 6-43 shows a circuit composed of three solar cells. All three are connected in parallel with each other. Notice that the positive terminals of all three are tied together. The same is true of the negative terminals. Wired in this manner, the output voltage will still be 0.45 Vdc, but the available current will be triple that of a single cell, or 0.3 ampere.

Now, what about the power output from this parallel circuit? Using $P = IE$ for power again, we find that 0.45 times 0.3 still equals about one-eighth of a watt. This is the same power which was derived from the three cells in a series connection.

A lesson should have been learned by the examples given here. Each cell or battery is capable of delivering a specific amount of power. There is no way to increase this power. Additional cells may be added. With each cell, the power increases. Three cells can deliver three times the power of one. Thirty cells can deliver thirty times the power of one cell, and so forth. The power is more or less fixed by the number of components in the circuit, but the manner in which the power is delivered to the load may be changed and is dependent upon the method by which the many cells are wired in respect to each other and to the load.

This may be a bit confusing to some readers. To put it more simply, the power will always add with each cell in the circuit. If one cell delivers one watt, then 20 cells will deliver 20 watts, regardless of whether the cells are connected in series or in parallel. In series connections, the voltage output of each unit is added, so 20 cells in series will deliver 20 times the voltage output of a single cell. Current remains the same and never adds in a series circuit. In a parallel circuit, 20 cells will deliver 20 times the current of one cell, but the voltage never adds.

We cannot change the total power availability of a fixed number of cells, regardless of the manner in which we connect them. We can determine if the power available can be correctly used by the circuit under power by the manner in which the cells are connected. An electronic circuit may require that power be delivered at 1.35 volts in order for it to be used. We can see that the power is delivered in a usable manner by delivering it at 1.35 volts through a series connection. Another circuit may require its power be delivered at 0.45 volts and at 0.3 amperes of current. This delivery can be most efficiently made by a parallel connection. Both circuits have the same power capability. The trick is to deliver it in a way which can be used. The voltage and current rating of the solar cell supply will determine this last factor, and the manner in which the cells are connected will determine the voltage and current ratings.

To take the discussion one step further, let's assume that our fictional electronic circuit needs power from the solar supply delivered at a value of 0.3 amperes at 1.35 Vdc. How would this be accomplished? First, work the $P = IE$ formula. Inserting the given values, we find that 0.3 times 1.35 works out to about three times the power rating of three of our sample solar cells. We already know that we cannot increase power without increasing the number of cells, so we must triple the number of cells to nine. This would give us approximately 0.4 watts of power, which is exactly what we need. But how do we combine the cells to equal 1.35 volts at 0.3 amperes? If we put all nine in series, then the resultant output voltage would be 9 times 0.45, or 4.05 Vdc. The circuit requires only 1.35 Vdc. On the other hand, if we combine all nine cells in parallel, we will have an output voltage of only 0.45 Vdc with a current of 9 times 0.1, or 0.9 amperes.

The answer to this riddle is a series-parallel circuit. This is shown schematically in Fig. 6-44. Three sets of three cells wired in series are then wired in parallel configurations. The result is three

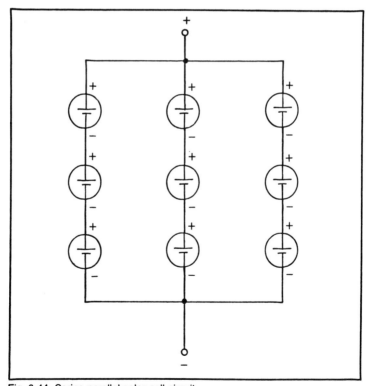

Fig. 6-44. Series-parallel solar cell circuit.

sections, each delivering 1.35 Vdc at 0.1 amperes in the series configurations. These might be called series blocks. When these blocks are connected with each other in parallel, the current adds and we finally arrive at a power supply which delivers a total output of 0.3 amperes at 1.35 volts. This is exactly what the circuit required.

If the circuit needed higher voltage, then more cells would have to be added in series to increase the output. In doing this, the current capacity of the power supply would remain the same, because cells are only added in series to the tops of the three series blocks. If the supply required 1.35 volts at a higher current level of, let's say, 0.4 amperes, then another string of three series cells would be added in parallel. Figure 6-45 shows several different solar cell arrangements using the basic cell in this discussion. Note the different output combinations which may be arrived at by adding cells.

This project involves building a solar power supply which is relatively inexpensive and can be used to charge a number of 1.5 volt rechargeable batteries. The number of cells which can be

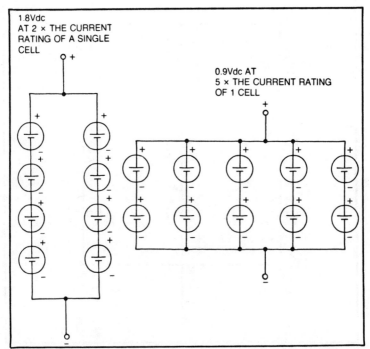

Fig. 6-45. Solar cells may be used in many different wiring configurations.

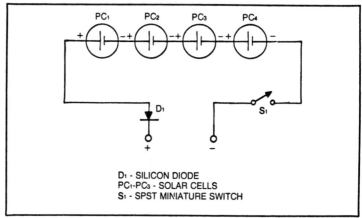

Fig. 6-46. Solar cell power supply/battery charger.

charged at one time will depend upon the amount of sunlight present on the surface of the four cells and the current rating of each individual cell used for this project. Today, many good buys can be found through surplus houses and even from your electronics hobby store when shopping for solar cells. No specific cell is designated with this project. Its choice is up to you. The cheapest cells will produce minimal amounts of current (20 milliamperes or so). This will be adequate to slowly charge one or two AA rechargeable batteries. On the other hand, you may elect to choose high current cells which may offer anywhere from .500 mA to 1.2 ampere output. These may be used for charging a number of D-type batteries at a fairly rapid rate. Also, the larger cells will provide reasonable output current when sunlight is diffused by light clouding. These may also be used to produce a light charge by subjecting their surfaces to artificial light present in a room or elsewhere.

The sole determining factors in solar cell choice will be cost and intended application. If you just want a simple supply which can be used to keep a few AA batteries charged, you can probably opt for the cheapest solar cells. On the other hand, if you want a circuit that you can really depend upon and which can be used to charge anything from a single AA cell to a whole bank of D cells, then you will want to purchase the largest solar cells you can afford.

Figure 6-46 shows the schematic diagram of our solar cell power supply/battery charger. It is made up of four solar cells wired in series. Notice that the negative terminal of the first is connected to the positive terminal of the second. The second negative terminal is connected to the positive terminal of the third, and so on. This series connection of four solar cells results in an approximate output voltage of 4 × 0.45 volt, or 1.8 Vdc. This is slightly higher than the 1.5 Vdc we are shooting for here, but a series diode is placed within the circuit to prevent damage to the solar cell bank should overcharging occur and the battery feed back through the cells. The insertion of this diode brings about a 0.3 volt drop. This leaves us with a nominal 1.5 Vdc. Actually, most rechargeable batteries deliver slightly less than 1.5 Vdc, so you could remove D1 from the circuit and use only three solar cells, which would deliver 1.35 volts dc. This happens to be the value of most rechargeable batteries, so a full charge could still be had. The deletion of one solar cell decreases the cost of this project by nearly 25 percent, but you will have to be careful during the operation of this supply during the charging mode to make sure damage does not occur to the solar cell string. I recommend the four-cell project because of the built-in protection offered by the diode. Using the diode with a three-cell string would reduce the output voltage to about 1.05 Vdc. This is too low for adequate charging.

This circuit is extremely simple and its actual construction will depend upon the type of solar cells which are used. The smaller types are often rectangular and can be easily mounted on a small section of perforated circuit board. The larger cells might be mounted on stiff aluminum sheeting which has been fitted with an appropriate insulating surface. If you use a metal mounting platform, make certain the insulator also provides some thermal protection. Heat is an enemy of solar cells and their performance will drop off as their operating temperatures increase.

Whatever type of mounting base you see for your solar cell supply, make certain that it is fairly rigid. If the surface is allowed to bend, the delicate solar cells may be mechanically stressed and will break. Most cells are constructed around an extremely thin sheet of glass-like material and will snap under even the smallest opposing pressure. If this happens, the cell is not destroyed. You can simply glue or tape it back together and use a small strip of foil to electrically bond the back surfaces (positive) to each other. Also, your mounting platform should be portable and equipped with some kind of leg or stand which will allow it to be angled toward the sun in such a manner as to catch the light rays directly.

Wiring is non-critical, although it is mandatory that the cells be connected in regard to polarity, as shown in the schematic drawing. The only nonpolarized device in this power supply is S1, which is a single-pole single-throw miniature toggle switch. Many builders elect to exclude this component from the power supply. This is certainly a practical idea, since no current will flow if a load is not connected to the positive and negative output terminals. For most applications, this load will be one or more rechargeable batteries.

When connecting the series diode, again, polarity must be observed. If you reverse this diode, no damage will occur, but your supply will deliver no output whatsoever. The diode should be a silicone rectifier type which is capable of passing the maximum current your supply can deliver. If each solar cell is rated to deliver 20 milliamperes maximum, then this will be the maximum current delivered by the four cells in series. Therefore, a silicon diode with a rating higher than 20 milliamperes will suffice. From a practical standpoint, this means that almost any rectifier will do, since all of the common types are rated for at least 50 peak inverse volts and forward currents of 100 milliamperes or greater. The garden variety of cheap diodes is usually rated at 50 PIV and 1 ampere. This will be adequate for nearly any solar cell types used, although there are a couple which will produce an output of greater than 1 ampere. These are very expensive and probably will not be selected for this project. But if they are, play it safe and use a 50 PIV diode rated at 20 percent more than the highest possible current from a single cell.

You do not have to supply a box or cover for mounting this supply. Obviously, the sensitized surface of each cell must be directed toward the sun or another source of light. Most solar cell power supplies use "open-frame" construction, and this one is no exception.

Many builders elect to use epoxy cement to secure the solar cells to their mounting platform. This is a good idea but must be considered permanent as far as the cells are concerned. It is nearly impossible to remove them without breakage after the bonding compound has set. For this reason, I recommend that you make all solder connections and completely check out the operation of this supply before permanently affixing the solar cells in place. If the bonding compound is applied before testing, an inadvertently reversed solar cell might have to be broken away and replaced with a new one, increasing the cost of this project by around 25 percent.

Once the project seems to be complete, examine the connections at each solar cell, making certain that the negative lead of one is connected to the positive lead of the other, and so on. Also check to make certain that the positive output lead connects to the positive (bottom) surface of the solar cell and that the negative lead is brought off the solar cell at the opposite end of the string. Inspect the diode to make certain it is correctly placed in the circuit. If all appears well, connect the probes of a dc voltmeter to the output leads of the supply. Activate S1 with the cells exposed to a bright source of light. For test purposes, this can be an incandescent bulb or fluorescent light in your workshop. You should get a reading of approximately 1.5 Vdc. Depending upon the solar cell type and the diode used, this reading may fluctuate by a few tenths of a volt. That's all there is to it. A reading here is an indication that the supply is functioning normally, and it may now be used for charging batteries or directly powering any device which requires a 1.5 Vdc input.

If you get no reading at this point, then more than likely, you have reversed a solar cell or a diode. This can be quickly checked by placing the probes of the voltmeter directly across the solar cell string, bypassing D1 and S1. If you get no reading, check the voltage across each solar cell. All this assumes that the cell's sensitized surfaces are being exposed to bright light. Using this method, you will quickly identify a defective cell, which will have a 0 output voltage. Such an occurrence is very rare. If you read normal voltage across the string when bypassing D1 and S1, this is a sure indication that the series diode is either reversed or defective. There is a very slight possibility that S1 could be defective, but this is a one in a million occurrence.

Once your supply is operational, you may want to install an appropriate battery holder at the output terminals. This will allow you to snap the batteries into place for charging. Your local electronics hobby store probably stocks holders which are designed for one to four penlight cells. Others are available for size C and D cells.

This power supply/charger can be further improved upon by adding an internal ammeter to read current output. This will help to determine the charge rate of the supply. Figure 6-47 shows the proper placement point for the meter, which will have to be chosen based upon the current abilities of the solar cells used. Notice that the meter is a polarized device and must be connected as shown in the positive leg for proper operation. It may just as easily be placed in the opposite negative leg, but it will have to be reversed. Notice that an additional switch (S2) has been included with this metering circuit. The meter induces a very slight resistance into this circuit, which will drop the voltage slightly. Since the output voltage from this supply is very low, it is necessary to reduce losses wherever possible. When a reading is desired from the meter, the switch contacts are opened, which allows current to flow through the

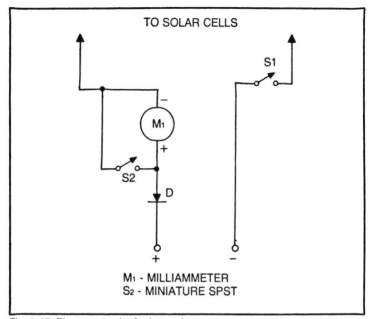

Fig. 6-47. Placement point for internal ammeter.

mechanism. Once a reading has been noted, the switch contacts are closed. This shorts out the meter, effectively removing it from the circuit, and a voltage drop is avoided during normal operation. You may be able to avoid using S2. The only way to find out is to insert it in your circuit and see what the output voltage reading is (under load) by using an external dc voltmeter. If the voltage is equal to or slightly higher than the dc value of the batteries being charged, then the shorting switch is unnecessary. If the charging voltage is slightly lower than desired, then the switch should be inserted to allow for a full charge potential.

The nice thing about this power supply is found in the fact that its operation costs absolutely nothing. The only expense involved is in purchasing the components. From this point on, all electricity delivered to the batteries being charged is free.

The total cost for this power supply can be anywhere from less than ten dollars to over $100, again, depending upon the solar cells chosen. Some of the cheapest types can be had for around one dollar, while some cost more than $40 each. Shop around, and you may be able to pick up some high current units at quite a cost savings. The author recently took advantage of a closeout sale at a local electronics hobby store. One line of solar cells was being discontinued and replaced by another. The cells were 500 milliampere units and normally sold for about nine dollars. They were purchased during this sale for less than five dollars each. You may be able to do as well or even better by checking as many commercial surplus outlets as possible. As of this writing, Edmund Scientific of Barrington, New Jersey is offering rectangular cells which deliver 22 milliamperes of current for $2.95 each. Heavy duty 1-ampere cells are approximately $14, and $18 will buy you one which will deliver nearly 2 amperes. This is only one source of solar cells but accurately reflects general market prices which prevail throughout the United States.

Some companies offer power supplies just like the one described here. They typically sell in completed form for about $90. These produce an output of approximately 400 milliamperes. By purchasing 400-milliampere solar cells and performing the assembly and wiring yourself, you can probably produce an equivalent unit for less than $30. This assumes, of course, that the mounting platform can be constructed from already available parts. This is quite a savings and indicates the economical aspects of building many of your own devices at home.

It was previously mentioned that heat buildup is an enemy of solar cell supplies. For this reason, make certain that your supply has adequate circulation, especially on extremely hot days. This power supply should not be placed in front of a closed window, as heat buildup here can be tremendous. Rather, it should be placed on the outside window sill, where it will catch the direct rays of the sun and still be subjected to cooling breezes. One exception to this rule is often practiced by the author in the summer when he places his solar power supply on top of his air conditioner, where the cold air is vented. The heat buildup of sunlight penetrating the plate glass window is completely removed by the steady stream of cool air traveling across the surfaces of the solar cells. Their cool operating temperatures assure maximum efficiency.

PROJECT 21
SURGE PROTECTION FOR MEDIUM TO HIGH VOLTAGE POWER SUPPLIES

Solid-state rectifiers are almost always used in power supplies built today. They far outweigh the old vacuum tube rectifiers in versatility, compactness and efficiency. However, solid-state devices in general are less immune to permanent damage due to extremely high instantaneous surges of current and voltage. When a power supply is initially activated, the filter capacitors are in a fully discharged state. The output from the rectifiers, then, is looking into the equivalent of a short circuit. This condition lasts for only a small fraction of a second until the filter begins to charge. At about a 50 percent charge rate, the power supply is operating normally. The old tube-type rectifiers were easily able to withstand this momentary surge. However, solid-state rectifiers can often be damaged by the surge current and the voltage spikes which can occur during the split-second period of initial activation. For this reason, it is necessary to protect these solid-state devices. This is especially true when a number of silicon rectifiers are used in series connections and in moderate to high voltage circuits.

There are several methods of preventing high initial surges upon power supply activation. One method involves the use of an adjustable transformer connected between the household main and the power supply transformer. The transformer acts as a rheostat and is used to slowly increase the primary feed voltage from 0 to optimum. This can be 115 of 230 Vac, depending upon the primary winding of the power transformer.

Using an extra transformer is quite expensive and adds to the overall complexity of the power supply project. A far simpler method is shown in Fig. 6-48. This protective circuit acts upon the

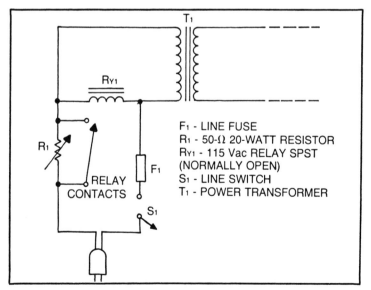

Fig. 6-48. Surge protection circuit for medium and high voltage power supplies.

229

principle of inserting a series resistance in the primary lead to limit the primary current flow and thus, the voltage seen across the primary winding. In other words, if this circuit is connected to a 115-volt line, the full 115 Vac will not be reflected across the primary due to the insertion of R1, the series resistor. However, it is obvious that R1 cannot continue to remain in the circuit, as the transformer secondary output voltage is directly dependent upon the primary voltage which, in this instance, should be 115 Vac. If the primary voltage is lower than nominal, this would be reflected in the secondary output and ultimately, at the dc output terminals.

Since surge damage occurs only during the first split-second of power supply activation, it can be seen that it is not necessary to leave the series resistor in the primary lead for longer than this period of time. The circuit shown here accomplishes the temporary insertion of R1 and then its complete removal from the primary circuit and does this automatically.

Referring to the schematic drawing, notice that RY1, a 115-Vac relay, is wired in parallel with the primary of the transformer. Its contacts are connected to either side of the series resistor. When RY1 is activated, its contacts close, shorting out the series resistor and effectively removing it from the primary circuit. All of this happens within an instant.

RY1 is an electromechanical component. When current flows through its windings, there is a split-second lag in its activation. This particular relay is designed to operate from 115 Vac, but due to the presence of R1 in the circuit, the coil does not see the full 115-volt potential. This further slows the relay from closing its contacts by a few milliseconds. The contacts finally close, however, shorting out R1; and a full 115-volt potential is seen across RY1 and the transformer primary. By the time RY1 closes its contacts, the filter capacitor in the secondary circuit of the power transformer has had time to partially charge at a substantially lower output than with a full primary supply, and the major part of the surge is avoided. When RY1 removes the series resistor from the circuit, the full potential allows the filter capacitor to be completely charged, and the power supply operates normally. When the supply is deactivated, RY1 opens its contacts again and the resistor is effectively placed back in the primary circuit, awaiting another period of activation.

This type of protective circuit is quite commonly seen in power supplies with medium to high voltage outputs. It can be effectively used with any power supply but is usually unnecessary when secondary output voltage (dc) is less than 1000 volts.

This project is designed to be added to any existing medium to high voltage power supply by simply adding RY1 and R1. If you are in the process of building a power supply, you can plan ahead to make room for the additional components.

The size of RY1 will be determined by the amount of current which is drawn in the primary leg. Ideally, the relay's SPST contacts should have the same current rating as S1. A good rule of thumb here is to rate both devices at 1½ times the maximum current which will be drawn through the primary. Therefore, if you expect a maximum of 6 amperes in the primary circuit, RY1 should have contacts that are rated to handle at least 9 amperes. Ten-ampere units are quite common, so this would probably be the type chosen. S1 and F1 would have similar current ratings.

R1 is specified as a 50-ohm, 20-watt resistor. This is a wire-wound type and is commonly available for a couple of dollars. This size is adequate for power supplies with outputs of up to 1000 watts dc. It will probably handle even bigger power supplies, because current passes through it for only an instant. Smaller wattage sizes may be used with equal results. This one assures a continued operation due to its being overrated. If this device should open up, no harm is done. This will most likely occur during an initial surge and the supply will simply refuse to operate (assuming that the relay contacts have not already closed). When this happens, R1 acts like a blown fuse, and no primary power is delivered to the transformer windings. If you experience repeated burn-outs of R1, you might want to replace it with a 50-watt unit. Alternately, this can be an indication of an extremely slow-acting relay, and this component might better be replaced. Incidentally, when you activate this power supply, you will not notice the delay in relay activation. The delay is measured in fractions of a second and is far removed from falling within our sensory range.

The installation of this circuit modification will vary from project to project. Some builders will elect to install a heavy-duty terminal strip on the power supply chassis, connecting the resistor and relay contacts here. If you do this, make certain the terminal strip contacts are rated to handle the power supply primary current. Alternately, the relay contacts themselves may serve as the mounting points for the resistor and the primary transformer and line cord leads. Make certain that S1 is installed between the line plug and the relay coil. Should this switch be installed at a point in the circuit past the relay coil, the line potential will be present at the relay at all times. You may still turn the power transformer on and off with the switch, but you will be getting no surge protection, since R1 will be continuously removed from the circuit.

Once this circuit has been installed, simply activate S1 and make certain your power supply is producing its normal output voltage at its dc terminals. It is good to visually examine the relay contacts to R1 to make certain they are removing the resistor from the circuit. Test your power supply under load. If you notice poor regulation properties or if R1 begins to heat and smoke, this is a sign that the relay contacts are not closing completely. This is probably due to a defective relay, and replacement will be required. If your dc output voltage is low, this is also an indication that R1 is remaining within the circuit. Another test method involves the placement of ac voltmeter probes on either side of the series resistor. When the supply is activated, no reading should be had at this point. If you read any amount of voltage, then R1 is still in the circuit and is not being shorted out by the relay contacts.

If you choose a good quality relay that is known to be operational and assure that R1 is operational as well, then it's almost impossible for this circuit not to work. This, of course, assumes that you began with a fully operation power supply at the start of this project. With this circuit installed in your medium to high voltage power supply, you will be constantly protected from rectifier burnout due to instantaneous surges upon activation. This circuit should serve you for the life of your power supply. However, if you notice a clattering sound coming from the contacts of the protective relay, this probably means that it's time this component be replaced. Should the relay become defective and open its contacts when the power supply is operating into a load which draws a high amount of current, the series resistor (having been placed back in the circuit due to the failure of RY1) will quickly be destroyed and the power supply will be deactivated.

PROJECT 22
HIGH VOLTAGE DC POWER SUPPLY

Many persons have need of dc power supplies which produce outputs in the high voltage range. Some technicians argue over just what high voltage is, but for the sake of this discussion, we will assume it's any value over 1500 Vdc. High voltage power supplies of between 1500 and 4000 Vdc are often used to supply operating potential to the plate circuit of radio frequency power amplifiers. Amateur radio operators are quite familiar with these amplifiers, as are those persons who work in commercial radio and television stations. The idea of a high voltage power supply reeks of complexity to those persons who are fairly new to electronic project building, but in fact, they are simpler than even the most basic low voltage supply which uses electronic regulation. By simpler, we mean theoretically and schematically. In practice, high voltage power supplies normally use larger components and sometimes it is necessary to combine a number of components in series to arrive at a circuit which will be adequate to withstand the high voltage ratings. This is especially true of rectifier and filter capacitor circuits.

Figure 6-49 shows a basic high voltage power supply circuit which is quite popular today. No component values are shown, since this is a theoretical circuit for discussion purposes only. Does it look familiar? It should, because it has been used several times in other projects in this book. It is a full-wave voltage doubler circuit composed of two rectifiers and two capacitors, and that's all there is to it. If the transformer secondary voltage is 1000 Vac, then the output at the dc terminals will be about 2.8 times this amount, or 2800 Vdc. This obviously is within the previous high voltage range of 1500 volts and above.

Before you begin to think that high voltage supplies are simpler to *build* than are many low voltage types with electronic regulation, look at our high voltage power supply project shown in Fig. 6-50. It

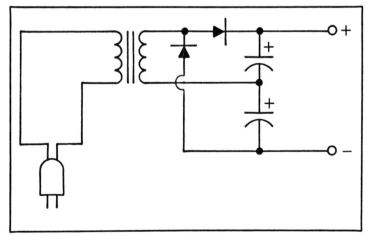

Fig. 6-49. High-voltage dc power supply circuit.

looks quite a bit more complex than the previous schematic, although they are both theoretically the same. Of course, the theoretical high voltage supply used a total of two rectifiers. This one uses eight. The former supply used only two capacitors, while this one uses eight. But to repeat, they are both operationally the same.

Before we move on to the power supply in Fig. 6-50, let's return to Fig. 6-49 and discuss component ratings. Each rectifier would have to have the ability to withstand 2.8 times the secondary ac potential. With the previous example, this would mean PIV ratings of at least 2800 volts, but this provides no safety margin whatsoever. Even a 3000 volt rating would not provide adequate assurances. No, each of these diodes should be rated at about 4000 PIV to assure a continued safe operational life in such a supply.

From a practical standpoint, single diodes with ratings above 1000 PIV at 1 ampere are very expensive. 4000 PIV units could easily cost $40 or more. A previous discussion stated that silicon diodes can be wired in series for increased PIV ratings. Obviously, it makes better economical sense to use four 1000 PIV diodes, costing about 25c apiece from industrial surplus channels, than it does to use a single unit which might cost forty times this much. When diodes are wired in series, however, it is necessary to install compensating resistors in parallel with each to assure proper matching. Additionally, parallel capacitors are used to prevent damage from voltage spikes. Fortunately, these protective devices cost very little, but it can be seen that the circuit has been made more complex due to a desire to be economical.

Now, let's move on to the filter capacitors. Since this is a voltage doubler circuit, each must be rated to withstand at least half of the maximum output voltage. In this case, the working voltage figure for each would be a maximum of 1400 Vdc. For most modern rf amplifier applications, good dynamic regulation is required, so the filter capacitor output should be at least 20 microfarads. In a full-wave voltage doubler circuit, the capacitors are in series with each other. Therefore, the total capacitance is decreased by the number of capacitors divided into the total capacitance of one component. If we use two capacitors, as shown in our theoretical circuit, then each would have to be rated at a working voltage of at least 1400 Vdc and with a capacitance of 40 microfarads each. As is the case with solid-state rectifiers, when capacitor voltage values fall into this high voltage range, prices soar. Then, too, it is practically impossible to find an electrolytic capacitor rated at higher than 450 Vdc.

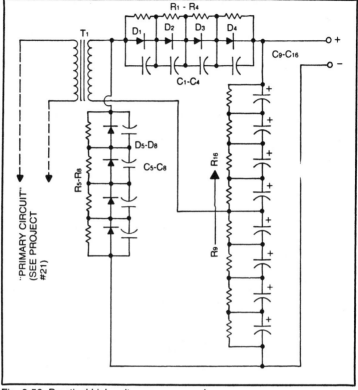

Fig. 6-50. Practical high-voltage power supply.

The circuit in Fig. 6-50 uses eight filter capacitors, each rated at 200 microfarads and 450 Vdc. This yields us an effective filter capacitance of 25 microfarads and a total working voltage rating of 3600 Vdc. Since the maximum power supply output will be on the order of 2800 volts, this is quite adequate and provides a good safety margin. Do you think that we could reduce the number of capacitors by one and still have a capacitor working voltage rating of 3150 Vdc? Not really. This seems fine until you consider that this is really two capacitor banks composed of four units each connected in series. Both of these banks have to be balanced. Therefore, if you remove a capacitor from one, you must do the same to the other. If two capacitors were removed, one from each series string, we would be left with a total of six 450 Vdc capacitors in series. This would increase our capacitance value to about 33 microfarads, which is no problem, but the working voltage rating of the series string has dropped to 2700 Vdc. This is below the maximum output rating of the supply and could cause the capacitors to short out. This can be disastrous, because if one shorts out, it means the rest must absorb the full voltage rating, which will be in excess of their maximum ratings. As more and more capacitors short out, more and more strain is placed upon those that remain. The result is all of those capacitors in the string are destroyed.

Referring again to Fig. 6-50, when a number of capacitors are used in a series string, it is necessary to equalize their internal resistances by wiring a high value resistor in parallel with each. In voltage doubler circuits, it is most convenient to make these high wattage units which can also serve as independent bleeder resistors for each capacitor. Our project specifies 25 kohm, 20-watt, wirewound resistors for each. These cannot be excluded because they not only assure that the filter capacitors are de-energized when the power supply is shut down, but make sure that one capacitor does not assume more than its fair share of the total voltage due to a lower internal resistance than the other units in the string.

This discussion has shown how a relatively simple power supply project can be made a bit more complex in high voltage applications. If we were not concerned with economics, the simplicity could be re-inserted into this circuit by using single rectifiers rated at 4000 PIV each. (This would do away with the matching resistors and capacitors.) Additionally, we could purchase only two capacitors, each rated at 1500 or more working volts with a capacitance of 80 microfarads. The cost advantage of building such a supply, however, would offer little incentive over a commercially manufactured supply with the same output potential.

For our project, Fig. 6-51 lists the components which will be needed to produce this high voltage power supply. The most expensive component can easily be the power transformer. While a single transformer is shown here, the author has successfully used two television transformers, each with secondary voltage ratings of 550 Vac. When the secondaries are wired in series, this produces a total secondary output of 1100 Vac. This can be directly substituted for

---

C_1-C_8 - 0.01 μF DISC 1000 VOLT
C_9 - C_{16} - 200μF 450 V ELECTROLYTIC
D_1-D_8 - 1000 PIV 1 AMPERE
R_1-R_8 - 470 kΩ ½ WATT CARBON
R_9-R_{16} 25kΩ 20 WATT (WIREWOUND)
T_1 - TRANSFORMER 1000-VOLT SECONDARY

Fig. 6-51. High-voltage power supply components list.

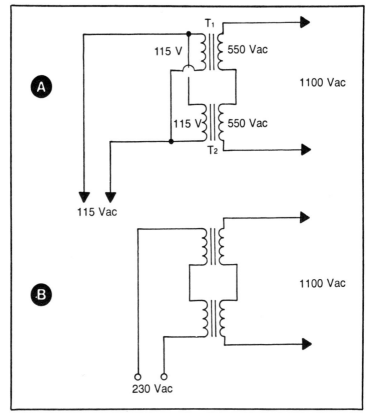

Fig. 6-52. Two-transformer wiring arrangement.

the 1000-volt transformer specified in the parts list. A slightly higher output voltage will be present, but this will still be within the ratings of the components specified here. Figure 6-52 shows the wiring arrangement using the two transformer system. Figure 6-52A shows the secondaries wired in series, while the primary windings are parallel-connected. The series connection at the secondaries allows for 1100 Vac output, but connecting the primaries in parallel means that full potential will be had from a 115-Vac line. For high power applications, many persons like to take advantage of the increased regulation efficiency that can be had from using a 230 Vac primary line. Using two transformers, each with a 115-volt primary, the arrangement shown in Fig. 6-52B will allow for a full 1100 Vac output with a 230-volt primary source.

If you will check through the many war surplus catalogs, undoubtedly you will come across a number of transformers which have secondary windings that produce an output of around 1000 Vac. Actually, any value up to about 1200 Vac is suitable and none of the components need be changed. Your maximum output voltage will be approximately 2.8 times the secondary potential. At a maximum of 1200 Vac at the secondary, the maximum dc output will be just less than 3400 volts. The series rectifier connections in this project are rated at 4000 PIV and the capacitor bank has a maximum working voltage of 3600 volts. If the transformer secondary should be higher than 1200 Vac, the resultant dc output might exceed the capacitor bank's rating. Many of the surplus transformers will be designed for operation from 115 and 230 Vac lines. Many of them contain dual primary windings that are wired in series for the 230 volt value and in parallel for 115 volts. This is exactly what we have done with the two-transformer circuits in Fig. 6-52. If you must buy a new commercially manufactured transformer with this output value, it can be rather expensive, costing $100 or more for 1000-watt models. The size of the transformer will depend upon the amount of power that is to be drawn from it and upon the type of application (continuous or intermittent). The author has found that you can usually double the power ratings of a military surplus transformer when it is used in intermittent duty applications and it will operate safely year after year. Commercial transformers designed for continuous service will also supply more power in intermittent duty and their ratings may safely be increased from 50 to 100 percent.

The components specified in this supply allow for a maximum safe current drain of approximately 750-800 milliamperes. This represents a power output of over 2000 watts, so an adequate safety margin has been built in.

Due to the heavy nature of the power transformer used for this type of supply, a heavy-duty chassis is an ideal mounting base. This may also serve as a support for the filter capacitor bank. Begin construction by mounting the transformer along with its specialized primary circuit, which is described in Project 21. Do not omit the special protective features of this primary circuit, because it is very easy to "pop" a silicon rectifier or two when the power supply is initially activated. Make certain that the transformer is firmly bolted to the chassis. Do not depend upon it to simply sit in place. While the transformer is quite heavy, any shift in the chassis could cause it to slide and destroy other expensive components within the power supply.

The next step involves the rectifier circuit. Ideally, this will be built on a medium-sized section of perforated circuit board. It is

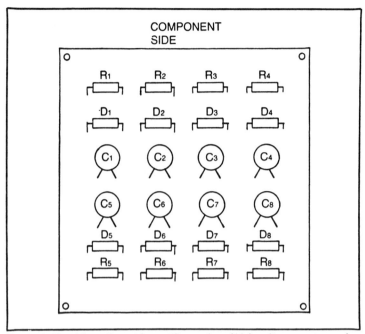

Fig. 6-53. Circuit board layout for rectifier portion of high-voltage power supply.

quite awkward to use a small terminal strip, as was done with a few other rectifier circuits, due to the fact that parallel capacitors and resistors must be included with each of the eight rectifiers. Figure 6-53 shows a suggested component layout pattern for the circuit board. This can vary to suit the builder's needs, but this particular pattern would appear to be the most likely for the power supply under construction. When dealing with high voltage power supplies, insulation becomes a big factor. Since high voltage is present at any point in the secondary circuit, including the rectifier assembly, the builder must be extremely cautious in order to prevent a short circuit or to place components so that the high voltage will cause a "flashover". The completed rectifier circuit board should be mounted on four bolts which may be pushed through the power supply chassis. Matching nuts will lock these bolts in place. One more nut is added to each shaft and is threaded down to about the halfway point. The circuit board, which has been drilled at each corner to accept these bolts, is then pushed down over each bolt and allowed to rest on the central nut. Four additional nuts are then threaded onto the bolts and tightened to the top surface of the circuit board. This procedure provides an extremely stable and inexpensive mount. Just make certain that the circuit board is slightly oversized for the components it must contain. Since each of the supporting bolts will be at ground potential, you don't want these coming in close contact with any portion of the rectifier circuit.

The next step is to wire the capacitor bank. It was stated earlier that it is necessary to isolate each capacitor in the filter string from ground and from the next capacitor in line. This means that the uninsulated metal case of each capacitor should not touch ground or the case of another capacitor. Some electrolytic capacitors may contain insulation around their cases. Be careful, however, because the insulation may be rated only as high as the capacitors' working voltage.

One trick I have used to provide a firm support for mounting a large number of filter capacitors is to cut a section of plexiglass into a rectangular pattern which is roughly three times the diameter of one capacitor in width. Its length will be determined by the number of capacitors, as well as their size. This assumes that capacitors with screw-in terminals are selected. This type of terminal arrangement is quite common on computer-grade electrolytic capacitors.

The next step is to drill holes in the plexiglass, through which the terminal screws may be inserted. This is shown in Fig. 6-54. This particular piece of plexiglass should be ¼" thick and is designed to support the eight electrolytic capacitors. After the drilling procedure is completed, mount the electrolytics as shown in the drawing by pushing the terminal screws through the top of the plexiglass and threading them into the capacitor terminals which are located beneath. The plexiglass insulator may be supported at each corner, as was the rectifier circuit board. This, of course, will require much longer bolts. Alternately, another section of plexiglass may be epoxied to the bottom of the capacitor bank and allowed to rest on the chassis. If this latter method is used, secure the bottom plexiglass section to the chassis with four small bolts.

The top plexiglass sheet makes an excellent wiring platform. It is quite simple to interconnect all of the capacitors as shown by preparing short lengths of stranded hookup wire to make the connections. While not shown in Fig. 6-54, each bleeder resistor is fitted with a lug to match the capacitor terminal at each of its leads and is simply wired in parallel with the capacitor. In other words, the capacitor terminals form the mounting base for the resistors, which are suspended above the plexiglass sheet by the stiffness of the hookup wire. Be absolutely certain that you observe polarity

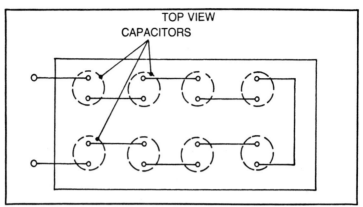

Fig. 6-54. Capacitor bank for high-voltage power supply.

while wiring this filter string. A reversal will not only cause inoperation, but possible capacitor damage as well. Remember, this is a series connection, so the negative terminal of the first capacitor is connected to the positive terminal of the next. This process is repeated throughout the string.

Once your capacitor bank has been properly mounted, it is a quite simple procedure to connect the output of the voltage doubler rectifier circuit as shown in the schematic drawing. Three connections are actually made. One rectifier output connects to the positive terminal of the top capacitor, another is connected to the negative terminal of the last capacitor in the string, and the third connection comes directly from the secondary lead of the transformer. High voltage output connectors are then installed through the power supply chassis. Later, I'll tell you how to fit this circuit with a tamper-proof metal cover, which will prevent persons from coming in contact with the high voltage elements.

At this point, take a long, hard look at the entire circuit. The interconnections can be a little confusing, even to professional builders, when the wiring process is taking place. After a brief respite from the tedium of building, a thorough inspection with the aid of the schematic diagram should bring any possible wiring errors to the front. Pay special attention to the polarities of each rectifier string, making certain that both are connected to the power transformer secondary and to the capacitor string as shown. Also, make certain that the polarity of each individual rectifier is correct, as indicated by the schematic. It is easy to see that a large number of component connections are required in the rectifier string, so be certain that each capacitor and resistor shunting each diode is correctly soldered in place. Examine the capacitor bank, again checking for any signs of polarity reversal. Make absolutely certain that each bleeder/matching resistor is firmly in place. Tighten any capacitor terminals that appear to be loose.

Re-check the primary circuit wiring. This is the high current portion of the power supply and may draw 20 amperes of current under high power operation. All wiring contacts must be tight, and all solder joints properly made. In some instances, it is best to avoid solder connections in the primary. Here, terminal strips of the screw-down variety which have large surface areas for current handling ability are used to make point-to-point connections. If you anticipate extremely high current in the primary circuit of your power supply, this might be the best way to go. Be sure that the fuse is of the correct size and that all components in the primary circuit are rated to handle the current which will be present here.

Testing this supply is very similar to the tests that were run on previous projects. First, connect a dc voltmeter across the output terminals. Using the components specified in the schematic drawing, the output voltage should be less than 3000 Vdc. If you substituted another transformer, this value could be higher. In any event, choose a voltmeter scale which is adequate to read the high voltage you are likely to encounter. One note of caution here: Many voltmeters contain high voltage scales which will measure 3000 volts or more. Many of these, however, (especially the inexpensive ones) may contain probe leads with insufficient insulation to provide adequate protection from these high voltage sources. Many voltmeter leads are rated for only 600 volts. For this reason, it is a wise idea to connect the probes to the power supply output in such a way that they do not have to be held in place by hand. When this is done, get away from the leads and the voltmeter and activate the power supply switch. You should get a reading approximately equal to 2.8 times the transformer secondary voltage. If, however, you notice the lights in the test area dimming, hear a loud hum or see an electrical arc, disengage the power switch and look for wiring errors. If you made an error in the rectifier circuitry, there is a good chance that many of the silicon diodes will have been destroyed. An arc is a sign of a short circuit, a terminal with high voltage potential in close proximity with ground, or a defective component which has given up the ghost. Caution: Once the power switch is deactivated due to a possible problem and the line cord removed from the wall, *do not* place your hands within the power supply circuitry until you have placed a shorting bar across each capacitor. A shorting bar (for these applications) will most often be the long metal shaft of an insulated screwdriver. Certainly, the bleeder resistors placed across each filter capacitor should drain the supply of all current. But should the power supply show signs of malfunctioning, anything could be wrong. It is quite possible that a wiring error may have effectively removed the bleeders from the circuit. Several bleeder resistors could also have been destroyed by a malfunction. Even though you may not get a voltage reading at the output of the power supply, this does not necessarily mean that the filter capacitors are not charged. There is no such thing as a slight electrical shock from a high voltage power supply. These are lethal voltages and can kill you quite instantaneously.

The previous checks discussed for problem solving in other projects in this chapter apply here. It is necessary to determine if primary current is being delivered to the transformer. Secondary output voltage must also be established. Then, check for output from the rectifier circuit, etc. This is such a simple power supply that troubleshooting should be no problem at all.

Be cautioned, however, because of the high voltage nature of this supply. Any checks run with a voltmeter should be done with the probes locked in place and with your hands completely away from them. This can be easily accomplished by fitting each probe tip with an appropriately sized alligator clip. Deactivate the power supply before removing these probes and be sure to bleed the capacitors by using the shorting bar after each power shutdown. This may seem ridiculous to some persons, but a discharged capacitor never killed anyone. Charged capacitors have.

Once your power supply has been made operational, do not neglect to fit it with an aluminum cover which cannot easily be removed by unauthorized persons. To leave a power supply which is capable of producing a high voltage potential in an open state is a gross act of negligence. Sure, should problems develop within the supply, it may take you a few minutes longer to gain access to the circuitry, but this is a small price to pay for safety.

The power supply is now ready to be used to supply plate potential to rf amplifiers or for any other application which requires a dc input of approximately 2500-2800 volts dc. You will find that this supply possesses good dynamic regulation due to the relatively large value of capacitance for a circuit of this type. The voltage will not swing terribly when going from a minimum current drain to maximum. This assumes, of course, that you have used the filter capacitor values specified. If it is possible to increase the capacitance of this circuit by using individual components with higher values, this will provide even better dynamic regulation. However, large value capacitors can be quite expensive and the units specified here are quite adequate for most high voltage applications.

PROJECT 23
ANOTHER HIGH VOLTAGE POWER SUPPLY

The previous power supply project resulted in a circuit which delivered approximately 2800 Vdc at its output. Many rf amplifier circuits require plate potentials around the 3000-volt mark. However, many others may require dc voltage of around 2000 Vdc. This latter value is quite easily obtainable by using commonly available components. As is the case with any high voltage supply, the power transformer is usually the most expensive item and sometimes, the most difficult item to obtain. We have overcome this problem, as is shown in Fig. 6-55, by using two commonly available transformers with primary and secondary windings connected in series-aiding. Note that this power supply is designed to be powered from the 230-volt line and uses a protective circuit consisting of RY1, a 230-Vac relay, and R1, a 100-ohm, 20-watt resistor. The operation of this primary protection circuit was described in an earlier project.

Transformers T1 and T2 are most commonly found in old console-types of black and white television receivers. You should be able to salvage two of these units from a couple of junked TVs, which may often be obtained at local television repair facilities. Most transformers in this category will offer a high voltage secondary winding of 750 to 800 Vac. These windings also contain a center tap which is not used in this application. Incidentally, the filament windings, which are not shown in the schematic, may be wired in parallel or in series to possibly obtain a heavy-duty filament supply for rf amplifier loads.

While this particular circuit shows the primary windings of the two transformers to be connected in series for 230-Vac operation, they may just as easily have been connected in parallel so that 115 Vac may be used. If you elect to do this, F1 should be changed to a 20-ampere fuse, RY1 should have a 115-Vac winding, and you can reduce R1 to a 50-ohm unit.

I built one of these supplies a few years ago using the largest power transformer I could obtain from junked television receivers. The finished power supply was used to power an rf amplifier which drew approximately 2000 watts. The current drain was often instan-

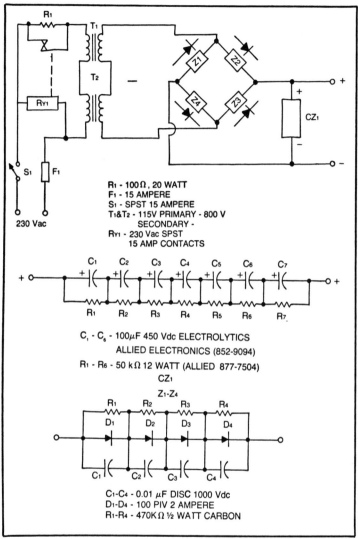

Fig. 6-55. High-voltage power supply using two power transformers.

taneous and certainly intermittent. The average power drain which the transformers were subjected to was about 500 watts. The transformers did get rather warm in these applications. However, the power supply was fully functional several years later when it was traded. As far as I know, it's still working perfectly today. Television receiver transformers are often designed to operate at relatively high temperatures. As long as there are no signs of smoke or burning, you are probably operating your supply within its limits. Then, too, the filament and other low voltage windings were not used at all in this supply. This increased the ability of the transformers to safely deliver this amount of power through the high voltage windings. Again, the author's use of the power supply discussed here involved intermittent duty. Yes, high amounts of power were drawn but only brief instances. There were much longer periods when very little current was drawn, allowing all components to cool. The same power supply certainly would not deliver a continuous 2000 watts of power without quickly destroying itself.

Referring to the schematic drawing, the series connection of the transformer secondaries yields an effective secondary ac potential of 1600 volts (1500 volts if the high voltage secondary winding of one transformer is 750 volts instead of the specified 800). This is rectified by a full-wave bridge circuit consisting of four complex minicircuits. These are shown in the inset of Fig. 6-55. Each of the four legs of the rectifier circuit consist of four 2-ampere diodes rated at 1000 PIV each. These are wired in series and are balanced by parallel resistors and protected from transient damage by disc ceramic capacitors, also wired in parallel. Again, the inset shows only one of the rectifier legs. Four of these circuits must be built in order to complete the full-wave bridge.

The rectified output from the power transformer is then filtered by a complex series capacitor bank. This is represented in the schematic by a CZ1. The complex breakdown is shown in the other inset of Fig. 6-55. Inexpensive 100-microfarad tubular electrolytic capacitors are used in this series string. Each is rated at 450 volts dc, and in this series connection, six of them provide a single effective capacitor rated at 2700 Vdc with a capacitance rating of just under 17 microfarads. This value is on the low side but is still adequate to give good dynamic regulation when used with any but the most demanding loads. If you'd like a little better regulation, 150-microfarad units may be opted for. These will cost more but will yield a total effective capacitance in this series string of 25 microfarads.

In order to balance the capacitors, assuring that each assumes 1/6 of the total voltage, six resistors are paralleled across each component. These are 50-kohm, 12-watt models which are easily obtainable. The actual ohmic value is not highly critical, but it is necessary that every resistor in this string be identical to the others. Slightly higher resistance values may be chosen. This will allow lower wattage devices to be used. However, the stored energy in the capacitors will not be drained quite so quickly.

A rather large, heavy-duty chassis will be required in building this power supply. The two transformers should be mounted together but not in extremely close proximity. Allow an inch or two gap between the transformer cases for air to circulate and remove heat efficiently. Make certain that when combining the transformers, the secondary (and primary) windings are wired in series-aiding. If the transformers you obtain are not exactly identical, you may have to experiment to see which series connection results in the desired total secondary output voltage. If a series connection results in a zero or greatly lowered output voltage, then simply reverse the connection for series-aiding.

Install the primary line circuit, along with the protective components (RY1 and R1). This supply is designed to deliver a large amount of output power (2000 watts peak), and a large amount of current will flow in the primary circuit when the power supply is delivering maximum output. For this reason, the higher 230-Vac primary potential is recommended. At 230 volts, approximately 9 amperes will flow when the supply is delivering 2000 watts. At the lower 115-Vac, the primary circuit will draw twice this amount to deliver the same power output. There will be less voltage drop in the primary circuit when using 230 Vac due to the lower current drain. Therefore, better overall regulation will result when using the higher voltage.

Once the primary circuit has been correctly wired, a large section of perforated circuit board should be chosen on which to mount the rectifier circuit. The previous project depicted a potential layout for a complex voltage doubler circuit. The full-wave bridge circuit used in this project is very similar to this previous one, except twice as many diode strings are used. You may refer to this previous drawing to see how the rectifiers should be configured on the circuit board, but double the board size in order to contain the extra rectifier strings. Be certain that you connect each of your completed strings in the standard bridge configuration. This is illustrated by the single diode symbols next to Z1 through Z4 in the schematic drawing. A reversal of any diode string (or for that matter, any diode within a single string) will result in the power supply not operating properly. A little extra care and patience at this stage of construction may prevent many long hours of removing improperly connected components and resoldering them into the circuit.

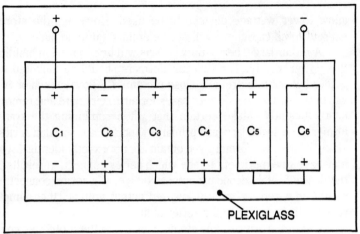

Fig. 6-56. The capacitor bank is constructed on a rectangular section of plexiglass.

Next comes the capacitor bank, which is best constructed on a rectangular section of plexiglass material, as shown in Fig. 6-56. The particular capacitors specified for this project have axial leads and are best suited to horizontal mounting. Notice the series connection, which is had by connecting the negative terminal of one capacitor to the positive terminal of the other, etc. The result is a six-capacitor string which has a common positive and negative input. Each capacitor may be held in place with epoxy cement or any of a number of other glues or commercial bonding compounds. Once the capacitor string has been completed, parallel each with its bleeder/matching resistor, and this portion of the project is complete.

Mounting the rectifier circuit board and capacitor board is accomplished by drilling holes in each of the four corners and using four through-the-chassis bolts as support legs for each. Make certain that the bolts which are at ground potential are kept an inch or so from any high voltage contact points. Connect the positive terminal of the capacitor bank to the positive rectifier output. The negative rectifier output is connected to the remaining capacitor bank terminal. Once this connection is made, the negative portion of the circuit may be attached to the chassis, which will serve as ground. The installation of appropriate high voltage output terminals completes this high voltage power supply project.

Checkout is accomplished after the circuit has been given a thorough visual inspection while referring to the schematic drawing. Be especially watchful for loose contacts or poor solder joints in the primary circuit. A high resistance connection here can cause arcing and complete inoperation. Also, check for short circuits in both the primary and secondary portions of the power supply.

With an appropriate dc voltmeter connected across the output terminals, activating the supply should produce a reading of about 2200 volts. This value will be closer to 2100 volts if the transformer secondaries were rated at 750 Vac instead of the higher value specified in the parts list. The line voltage can play a big part here as well. Throughout this book, we have referred to line voltage as being a value of 115 or 230 Vac. In some areas, the values will be closer to 110-220, while in others, they will rise to 120-240. If your area receives power at this higher value, then the output from the power supply will be slightly higher. Likewise, lower line voltages will result in a lower dc output. Some commercial and military surplus transformers have a tapped primary winding, which can set up to give optimum secondary output at any value from 95 to 260 volts. These transformers are quite convenient when high voltage output is quite critical. For most applications, however, a hundred volts or so at the dc output can be added or subtracted without any hardship.

If your power supply refuses to operate, deactivate the primary and *remove the line cord from its receptacle*. Take a long screwdriver with an insulated grip and short out the contact of each filter capacitor. If the usual checks for faulty wiring, component reversal, etc. turn up nothing, it will probably be necessary to run a value check from the line cord to the dc output. This must be done with the supply under power and possibly generating a potentially lethal voltage. Even a slight momentary brush with a high voltage contact point can cause death, so proceed with caution. With the supply activated, read the potential across the transformer primaries with an ac voltmeter. You should read the equivalent of the ac line voltage in your area. Next, check the secondary output with an ac voltmeter capable of reading 200 Vac or more. Don't depend upon the insulation at the voltmeter probes to necessarily protect you from this potential. Play it safe by deactivating the supply, fixing the probes to the secondary, standing back and activating the supply once again. Deactivate the supply before removing the probes. If the secondary output seems to be satisfactory, temporarily disconnect CZ1 from the circuit. A dc voltmeter placed across the positive and negative output leads from the rectifier circuit should produce a pulsating high voltage output of approximately 1500 to 1600 volts dc. If you get nothing here, then a problem exists in your rectifier circuit (most likely, a damaged or reversed rectifier), and repairs or replacements must be made. If the output appears to be satisfactory, then a problem exists within the capacitor bank, or there is a broken lead between the bank and the output terminals.

If the line fuse should open (often accompanied by room lights dimming and a vibration or hum), this is a sign of a major wiring error or a short circuit. If this happens, replace the fuse but do not attempt to reactivate the supply until the error has been found and corrected. If the short exists near the dc output terminals, the bridge rectifier circuit may be destroyed.

As is always the case with high voltage power supplies, some means must be taken to prevent accidental contact from persons who do not realize the dangers of high voltage. A sturdy aluminum case may completely cover the circuit. This supply should not be readily accessible without removal of the external cover. As a final operation check, activate the supply, note the output reading, and then flip off the switch. Watch the voltmeter to make certain that the stored potential in the capacitors decays to zero within a short period of time. If the charge continues to persist, there may be a defect in the bleeder resistor circuit. This should be corrected before the power supply is allowed to be put into service.

I have found that black and white television transformers can provide a great deal of power output in intermittent duty cycles. This supply was designed for such a use. However, if you need more power output, this may be obtainable by installing a heavy-duty blower in the power supply in such a manner as to direct cooling air over the cases of the transformers. The air should be vented through a shielded exhaust port consisting of a number of holes drilled in the side wall directly in line with the blower and transformer. These holes should not be large enough to allow for the passage of human fingers, especially children's size human fingers. By increasing the ability of the transformers to get rid of heat, the power output may be increased by 10 to 20 percent (even more in some instances and duty cycles) without actually exceeding the transformers' normally-cooled ratings. Excessive sudden voltage drops are a good indication of output current having exceeded the ability of the transformer.

At this point, it might be appropriate to talk a bit more about transformers, especially those that might be used in the last two high voltage power supply projects. While old black and white television consoles are still quite abundant as junk that must be hauled away every day by television repair shop personnel, persons in some areas may have difficulty finding these. The two-transformer arrangement in this project is economical only if junk or surplus transformers can be obtained. If you can't locate an old television receiver from which a transformer can most likely be had for free, your next bet is the war surplus market, where many different types of transformers can be had for a few dollars. Military ratings are greatly understated, and it's quite common to operate these transformers at double their power output without damaging the device.

But let's assume that you really want to build a supply using surplus television components. If you can't find a junk black and white console, you can probably find a large number of old color television receivers whose picture tubes have died. When this happens, most older sets are simply hauled away. Persons are usually informed that their television set is not worth fixing by the repairman, who usually gets struck with having to discard it. Color television transformers are not as ideal as those found in black and white sets. Their secondary high voltage windings may only produce an output of 200 volts or so, although some will closely match the values of black and white television transformers. You really won't know what you're getting until you've removed it from the chassis and tested it out. If you can't find a transformer (or in this case, two transformers) which come close to matching what is specified in the project parts list, improvise. For example, four transformers, each with 200-Vac secondaries, can be combined with secondaries in series to deliver 800 Vac with a power output equal to four times that of a single transformer. The project just discussed would require eight of these transformers with secondaries wired in series-aiding to produce the desired 1600 Vac. Obviously, an eight-transformer power supply would fill a tremendous space. However, four could probably be accommodated without too much of a size increase. The 800-volt combined secondary could then feed a full-wave voltage doubler circuit and you would still end up with a dc output of about 2200 Vac. This is known as making do with what you have on hand. The power supply in Fig. 6-55 used a full-wave bridge rectifier, which was desirable because it would deliver about 1.4 times the total secondary voltage of the two transformers wired in series. But we could just as easily wire the secondaries in parallel (for the increased power handling capability) and use the voltage doubler circuit to arrive at the same output. There are no distinct advantages to using either rectifier configuration in most cases, so it's left up to you as to just how you want to proceed. The purpose of all of these projects is to give you many different ideas on how power supplies can be constructed. You will find, in most instances, that a desired output voltage and current may be achieved by many different means and circuits.

PROJECT 24
DC TO DC POWER SUPPLY

A dc to dc power supply is one which requires a dc voltage input of one value and produces a dc voltage output of another. The great majority of dc to dc power supplies are designed to operate at inputs of approximately 12 Vdc. More accurately, this value can be said to be between 12 and 14 Vdc, since most automotive electrical systems offer outputs of about 13.5 Vdc. Power supplies of this nature are usually designed to be operated in automobiles. Their outputs often supply working voltage for communications devices such as transmitters and receivers, public address systems, etc. These are often tube-type devices and will require some medium to high voltage potentials for operation.

You have already learned that transformers are alternating current devices. With dc inputs, high current will flow in the primary windings and the device will be quickly destroyed. Power supplies with dc inputs and outputs operate in a different manner than ac-derived transformers. The dc input must be changed to alternating current in order to be fed to the primary of a transformer. From this point on, operation is very similar to that of an ac-derived power supply. The transformer secondary output is rectified and filtered to arrive at a final output voltage.

Figure 6-57 shows the schematic diagram for a simple dc to dc power supply which operates from a nominal 12 Vdc input and produces two output voltages. The high value is at 650 Vdc, while the low value is ½ this amount. These potentials should be adequate to supply operating voltage to many different types of vacuum tube transceivers. Maximum current drain from each supply leg should be kept to within approximately 225 milliamperes.

The heart of this power supply is a special transformer which is designed to be used with these types of circuits. The primary offers two windings. Looking at the schematic, the lower winding is the primary, while the top winding provides feedback to the bases of Q1 and Q2. This causes them to oscillate and switch on and off. This on and off switching controls the dc input voltage in such a manner that

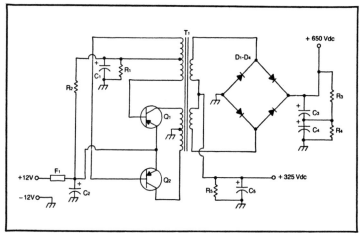

Fig. 6-57. Dc to dc power supply.

a simulation of alternating current is supplied at the primary winding. This is not a true ac voltage but a square wave signal. In any event, the constant variance of the voltage allows the transformer to operate much the same as any other. The transformer is very small and lightweight, because the polarity reversal of the square wave input to the primary is much faster than standard ac house current. The latter has a frequency of 60 hertz (60 full cycles in one second). This power supply has a frequency of over 1000 hertz. The faster switching rate allows the transformer construction to be centered around a lightweight, powdered iron core. While being significantly smaller and lighter in weight than most standard 60-hertz transformers, dc to dc power supply transformers can deliver relatively high amounts of power. For example, this power supply weighs only a few pounds and delivers approximately 200 watts of power output. Figure 6-58 provides a complete list of components that are to be used with this supply. You will notice that the secondary components are those that have been commonly used with other power supplies in this chapter which offer similar output voltages. The input circuit is the one that is quite different from the others we have discussed.

This power supply can be mounted in an aluminum compartment. It will be necessary to attach a rather large heatsink, as during full power operation, the transistors will be conducting nearly 20 amperes of current. Heatsinks designed for dc to dc power supply operation can sometimes be obtained from a local hobby store, but you may have to resort to a commercial outlet to find those sinks

C_1 - 10 μF 50 Vdc
C_2 - 10 μF 25 Vdc
C_3 - C_4 - C_5 100 μF 450 Vdc
D_1-D_4 - 1000 PIV 1 AMPERE
F_1 - 25 AMPERE
Q_1 - Q_2 - ECG105 (SYLVANIA)
R_1 -
R_2 -
R_3, R_4 - 25kΩ 5 WATT
R_5 - 50 kΩ 5 WATT
T_1 - TRIAD TY84

Fig. 6-58. Components list for dc to dc supply.

which offer enough surface area to really be effective. Begin by mounting the transistors to the heatsink using the mounting hardware which usually comes with each solid-state component. This consists of an insulating washer which is placed between each transistor and the heatsink surface. The heatsink, the washer and the transistor should all be given a liberal coat of silicone grease to assure thermal conductivity. Once the heatsink has been fitted with the transistors, wire leads should be attached to the three electrode points on each device. This assembly may now be set aside while the remainder of the project is completed.

Next, mount the transformer in its aluminum enclosure and connect the secondary circuit components first. Properly identify the leads which attach to the two secondary windings and connect them in series, as shown in the schematic drawing. The next step involves connecting the full-wave bridge circuit to the output of the secondary. Some space can be saved by using an IC-type of rectifier package. If not, mount the rectifiers on a tiny section of perforated circuit board. This is an economy power supply which derives its moderate voltage output from the center tap of the series connected secondary. Filter capacitors C3, C4 and C5 are 100-microfarad units. Two are combined in series in order provide a high enough working voltage rating in the high voltage filter circuit. Here, an effective capacitance of 50 microfarad is provided. This is quite adequate for good dynamic regulation. A single 100-microfarad capacitor is used in filtering the moderate voltage output leg. Notice that bleeder/equalizing resistors are installed in parallel with C3 and C4. A single bleeder resistor, R5, assures that C5 is completely discharged when the power supply is deactivated.

Once the secondary portion of the circuit has been wired, begin work on the primary. Start by identifying the various leads. The manufacturers' data sheets should answer any questions you may have. Ground the center tap of the primary winding (lower). Connect the two outside leads of the feedback winding to the bases of Q1 and Q2. This is best accomplished by mounting the heatsink to the enclosure and connecting the leads that were brought off each base electrode to the appropriate tap on the transformer. Since only a small amount of current will flow in the base circuit, standard electronic hookup wiring can be used here. However, the emitter and collector circuit wiring must use heavy conductor, because nearly 20 amperes will flow here during peak operation. The same is true of the 12-volt lead which connects to the emitter circuit. When building this supply, the author used #12 stranded wire for these portions of the circuit. Alternately, you may wish to combine several parallel conductors of #16 stranded wire. If this wire is not adequate, severe voltage drops will be incurred and the power supply will function erratically or not at all.

A 25-ampere fuse protects the circuit from severe overload, and capacitor C2 helps to stabilize the input voltage should mild fluctuations occur due to high current drains from other devices operated from the same electrical system. No switch is shown in this schematic drawing. Often, these types of power supplies are switched in and out of the electrical system by means of a separate relay. Since a maximum of about 20 amperes will be drawn by this circuit, it would be possible to install a large switch with contact ratings slightly higher than this in the positive voltage leg. Switches with contacts rated to handle this amount of current can be quite expensive and overly large for a compact mobile installation. In the long run, it may be best to use a relay which will operate from 12 Vdc with contacts rated to handle about 25 amperes.

To test this circuit, connect the input to a 12 Vdc source. Make certain the source is capable of supplying the current which will be drawn under load. As soon as power is connected, you should hear a whine of reasonably high pitch coming from the supply. This will be the switching transistors breaking into oscillation. If you do not hear the whine, recheck your primary circuitry, as a wiring error has probably been made. Hopefully, your precheck visual inspection will have uncovered any errors, because the switching transistors can be quickly destroyed from failure to break into operation.

With the supply properly functioning, you should read an output of approximately 650 Vdc under load at the high voltage output terminal. The medium voltage output should be measured at approximately 325 Vdc. If the output voltage is far lower than this, the primary circuit wiring in the high current leg may be too small or the hookup wiring between the 12-volt power source and the power supply is inadequate to handle the current. Because of the low input voltage, drive current is usually much higher than in most ac-derived power supplies, so one must be constantly aware of the severe voltage drops which can ensue. Remember, a drop of 3 volts is very insignificant at 115 Vac. At 12 Vdc, this represents a 25 percent decrease or the equivalent of a nearly 30-volt drop at standard ac line potentials. Even a drop of 1 volt at the input of this dc to dc power supply will cause the output voltage to drop by about 10 percent. Obviously, one must use short conductors within the wiring of the high current portion of the power supply and make sure that all solder joints are correctly made. At 20 amperes, a resistance of only ½ ohm in a poorly made solder joint will cause the voltage to drop to almost nothing.

Once your supply has been made operational, you may wish to install it permanently in your automobile. If this is done, make certain that it is not placed near a heater duct. The transistors will get very hot during normal operation and depend upon the heatsink to remove a great deal of this heat. Heatsinks accomplish their function by offering large finned surfaces which transfer the heat to the cooler atmosphere. Obviously, if a large volume of hot air is blowing across the heatsink from the automotive heater, the transistors will operate at far higher temperatures. This could easily cause their destruction, and these are not inexpensive items to replace.

The enclosure for this power supply should be made watertight. Mobile use of any electronic equipment can place quite a few strains on circuit components. During the winter months, this circuit may be cooled to near zero temperatures by remaining in the automobile overnight. The next morning when the automobile is driven, the heater is usually activated and the circuit temperature is brought up to a normal level very rapidly. This constant heating and cooling can take its toll, especially if the circuit is activated before being given the chance to warm slightly. We are speaking here only of temperature which might be encountered in very cold areas. Most solid-state components will operate normally from temperatures at a low of about 15° to a high of over 100°. However, the author does not like to operate dc power supplies of this type at full output current in cold weather until the circuitry has had time to reach normal temperatures. This can be done very rapidly by simply activating the power supply without drawing heavy current and allowing the idling current to warm the transistors for a few minutes. After this period of time, normal operation may be had.

Never abuse this power supply by drawing excessive amounts of current, even for brief instants. You can often get away with this with ac-derived power supplies, but remember that this project has two transistors in the primary which must conduct the full operating current. Drawing an extra 50 milliamperes at the high voltage output means the input transistors must conduct another 3 amperes or so. Transistors have a habit of self-destructing instantaneously when operated in excess of their maximum ratings, unlike such devices as transformers, which may overheat and smoke for several minutes before actually giving up the ghost. In other words, you have no warning that a transistor is about to go. It just quits.

PROJECT 25
12-VOLT INVERTER CIRCUIT

An inverter is a special type of power supply which accepts a dc input and produces an ac output, usually with a higher value. It is appropriate that an inverter type of power supply be discussed at this point in the chapter, because the previous project also contained an inverter circuit. In the previous discussion, it was mentioned that a dc to dc power supply needed to change the dc input to ac before applying it to the transformer primary. Once the dc had been changed to ac and placed at the primary, the secondary was able to produce its output. At this point, a rectifier and filter circuit was added to change the transformer ac back into dc again. If we had omitted the rectifier and filter circuit, then this dc to dc power supply would have been known as a dc to ac inverter, a circuit with a dc input and an ac output. This previous project, however, had a secondary output voltage of around 650 Vac. This is far too high a value to operate most of the ac devices we are accustomed to. Most inverter power supplies offer an ac output of the same value as house current that we use daily. Most are designed to accept a dc input of 12 to 14 Vdc so that they may be operated directly from the automotive electrical system.

What are inverters used for? Almost everything! Anytime you want to power an ac-only device from a battery, an inverter circuit is required. A few decades ago, inverters were electromechanical devices which depended on a vibrating reed contact to switch the polarity of the dc input voltage at a 60-hertz rate. Today, these mechanical vibrators are almost never seen, having been replaced by solid-state oscillator circuits. These have the advantage of being compact, highly efficient, and capable of switching rates which can be adjusted from very low values to frequencies of 15 kHz or more. The previous dc to dc power supply was set up to switch at a very high frequency when compared to normal ac line frequency. This was because the high switching rate allowed for the use of a compact, highly efficient transformer and because the ac output was to be converted back into dc. The high switching rate made filtering far more efficient as well.

But inverter circuits cannot take advantage of the efficiencies which can be had at high frequency switching rates, because their outputs must drive standard 115-Vac devices. Therefore, their power transformers are usually of the iron core variety and closely resemble standard power transformers. As a matter of fact, you can think of an inverter transformer as a device which accepts a 12-volt ac input and has a secondary which will produce approximately 115 Vac. It's almost like a 12-volt filament transformer in reverse. The inverter transformer does contain a feedback winding which allows the transistor multivibrator circuit to break into oscillation.

Figure 6-59 shows the basic power supply circuit, which closely resembles the previous project. Notice, however, that there

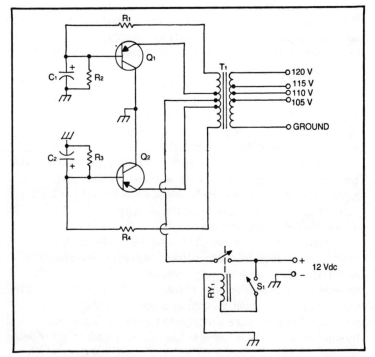

Fig. 6-59. 12-volt inverter circuit.

```
C₁&C₂ - 2.5µF 50 V
Q₁ Q₂ - ECG105 (SYLVANIA)
R₁-R₂ - 220Ω 25 WATT (WIREWOUND)
R_Y1 - 12-VOLT RELAY, 25-AMPERE CONTACTS
S₁ - SPST 1 AMPERE
T₁ - TRIAD TY75A
```

Fig. 6-60. Components list for inverter supply.

is no rectifier and filter circuit at the transformer secondary. As a matter of fact, there are no external components at all at the transformer secondary, which is used to directly drive the ac load. Figure 6-60 is a complete parts list. The only unusual component is the inverter transformer. This is a triad TY-75A which will deliver a steady output of 100 watts to an ac load. Intermittent loads of up to 150 watts may be powered for brief periods of time, but always observe the 100-watt continuous limitation.

Construction is quite simple and involves mounting the inverter transformer on a medium-sized chassis. It will be necessary to purchase a heavy-duty heatsink for the two transistors. This is also mounted to the same chassis. Mount the transistors to the heatsink before permanently attaching it to the chassis by inserting the leads through the appropriate holes. Naturally, an insulting washer will be fitted beneath each transistor to electrically isolate these devices from the heatsink, which is at ground potential. Do not neglect to apply heaping amounts of silicone grease to the heatsink surface, both sides of the insulating washers, and to the transistors themselves before bolting them down. Without the silicone grease, a good thermal contact may not be had. This will cause the transistor cases to heat beyond maximum temperatures, quickly destroying the delicate internal crystal.

Before bolting down the heatsink, bring leads off the transistor electrodes. Use #12 stranded wire for the emitter and collector portions of the circuit. Standard hookup wire may be used for the base circuits. When the heatsink is mounted, attach the emitters to the appropriate transformer taps and tie the collectors together, grounding them to the chassis. Next, the base circuits may be wired. Be sure to observe the polarity connections of C1 and C2. A reversal here will cause failure.

The wiring of the primary circuit is really quite simple once the transistors have been properly installed on their common heatsink. While the circuit may look a bit complex, there are only six components in the basic primary circuit in addition to the transistors. Once this portion of the circuit has been completed, the transformer center tap lead is connected to one of the contacts of relay RY1. The other contact connects directly to the positive 12-volt source. RY1 contains a 12 Vdc winding and is activated by throwing S1 to the on position. Since the coil does not draw a large amount of current, a simple 1-ampere switch of the SPST type is used. Alternately, a 25-ampere switch could be installed in the positive 12-volt line and the relay eliminated. No fuse is provided here, although it would be a simple matter to insert a 20-ampere slow-blow type in the positive line. Normally, these power supplies are connected to the automotive electrical system, where an in-line fuse may be used. Alternately, connections may be made at the fuse block, which provides automatic protection. This latter installation is not highly recommended, due to the high current drain of this supply. Most persons elect to bring the positive lead directly from the 12-volt battery terminal.

Examine the primary circuitry closely before attempting to check this power supply for operation. If an error exists, one or both power transistors may be destroyed. These are relatively expensive devices, and a thorough inspection will avoid this possibility entirely.

Notice that the transformer secondary has several taps. This allows for the selection of the line voltage which is most suitable for the devices to be powered. It also allows for voltage adjustment based upon the actual voltage of the input. For instance, if the input voltage is exactly 12 Vdc, primary circuit losses may drop this value a bit, and the output at the 115-volt secondary tap might be closer to 110 volts. Here, the 120-volt tap might actually be delivering 115 volts, which is what you're looking for. On the other hand, in an automobile electrical system with the alternator operational, the input voltage will be closer to 14 volts than 12. Here, the 115-volt tap may be producing 120 volts, so you would switch to the 110 or 105-volt tap to get the desired 115-volt output. This may sound a bit confusing, but remember that the printed output values on each tap are accurate only when the exact design voltage is used at the input. You may also find that some equipment will operate better at 120 volts than at 115. There is very little difference in most applications, but some specialized devices may provide better efficiency at slightly higher input voltages. Maximum current drain from the secondary of the supply will be less than 1.5 amperes, so it would be practical to install a four-position rotary switch which would manually select any of the four available secondary taps. You might even want to provide this supply with an ac voltmeter so that output voltage from this supply can be constantly monitored. If you elect to use this switching arrangement, choose a switch which is rated to handle at least 2 amperes of current. Do not switch taps while a load is being drawn, as this could damage the supply or even the device being powered. All switching should be done under no-load conditions.

Testing of this supply is very similar to those tests which were run on the previous project. Again, a close visual inspection is mandatory to avoid a potential component disaster due to a short circuit or wiring error. Since the transistorized multivibrator is switching at 60 hertz, you will not be able to hear the oscillation. For the initial test, use a 5-ampere line fuse and connect an ac voltmeter between one of the secondary taps and the bottom or ground lead of the transformer. Incidentally, while this lead is called the ground connection, it should not be connected to the chassis. This ground should be kept separate from chassis ground.

Connect a 25-watt light bulb to the output of T1. Activate the supply by switching S1 to the on position. You should hear the click of the relay (RY1) and immediately obtain a reading of approximately 115 Vac on the voltmeter. The actual reading will depend upon which tap you are measuring across. If nothing occurs, deactivate the supply and visually inspect the circuit again. If you find no errors, you will have to run a check with a dc voltmeter to determine if primary current is being delivered to the supply. If it is and there is no output, then there is a problem with Q1 or Q2 (possibly both) or a wiring error in the base circuit(s). Make certain that you have observed correct polarity when installing C1 and C2. Check Q1 and Q2 on a good transistor checker. If damage has occurred, go over the circuitry one more time to find the fault before replacement transistors are installed.

This is a very simple circuit and should work the first time if you have followed the schematic drawing accurately. You may now connect any of a number of ac devices to this inverter circuit for ac operation from a dc system. In order to protect your power supply and to assure proper operation of any equipment which derives power from it, a few rules must be followed. First of all, the power supply should not be activated until a load is attached. It is best to connect the device to be powered to the supply and then activate the supply. For example, if you were going to power a radio receiver from this supply, you would attach the receiver, turn it on and then turn on the power supply. Deactivation is the reverse of this. You would first turn off the power supply.

While this type of power supply will provide operating current to many different kinds of electric and electronic equipment, there are some devices which are simply not applicable to this type of power system. Naturally, any equipment which draws more than 100 watts continuously is too large for this supply. Some equipment, especially those with internal motors, may require vast amounts of starting current for a second or so and will then operate at low power levels well within the capabilities of this supply. However, these devices should not be attached to this type of supply, because the transistors can be damaged by the high initial starting current required. Remember, solid-state devices do not normally stand up well under temporary overloads.

Almost any type of motorized device, whether it draws high starting current or not, will not function properly with an inverter power supply. Many types of motors depend upon the frequency of the ac sine wave to establish speed. The output of this supply, while alternating in nature, is not a true sine wave. It is really a square wave and will play havoc with the speed of motorized devices. Additionally, some audio devices may require line filters to remove the interference caused by the constant switching of the input current. This will be heard as interfering clicks or pops at the output speakers.

Now that we have mentioned the kinds of devices which cannot be powered, we can look at those which are very applicable to this type of power supply. Almost any type of electronic equipment which uses a power transformer and does not draw current in excess of the previously discussed maximums will work well with this power supply. This includes radio transmitters and receivers, public address systems, electric lights, some appliances (non-motorized), and a myriad of other devices. This brings us to a specialized use for this inverter circuit.

The previous dc to dc power supply accepted a dc input and ended with a high voltage dc output. There is no reason why any of the other power supplies in this chapter which derived power from the ac line cannot be used with this inverter circuit. This, of course, excludes the high power ac-derived supplies, which would draw too much current. As shown, the inverter circuit is a dc to ac supply. By attaching an ac-derived power supply (ac to dc), you end up with a complex dc to dc supply. Actually, the steps taken here are identical to those taken in the previous project. Referring to the earlier project, a dc input was changed to ac, passed through a transformer, and then rectified for a dc output. Using an ac to dc supply with this inverter circuit, again, the input is dc. It is changed to ac by the two transistors and fed to the inverter transformer. The transformer output is connected to the primary circuit of the ac-derived power supply. Its secondary ac output is then rectified for a dc output. The only difference in the two supplies described here is that an extra transformer is used when an inverter circuit and an ac-derived power supply are operated together. This is just one of the many interesting aspects of using an inverter circuit.

To make the operation of this inverter a little more convenient, the circuit shown in Fig. 6-61 can easily be added to the secondary output of T1. S1 is a four-contact rotary switch which selects the desired transformer tap. This may be deleted if desired. The main portion of the circuitry involves connecting a standard female receptacle to the transformer output. The receptacle should be a type designed for chassis mounting, as it can be installed near the transformer secondary directly through the chassis surface. Alternately, a line cord can be installed with the female receptacle at one end. F1 is a 2-ampere, 115-Vac fuse which prevents the power supply from being severely overloaded. Another optional feature of this circuit is M1, which is an ac meter used to monitor the output voltage. It is installed in parallel with the output and will allow the tap settings to be adjusted for the desired operating voltage. The best way to accomplish this is to plug in the load, activate its on switch and then activate the inverter power supply. Note the voltage reading. If you desire to change it, switch off the power supply, adjust the rotary switch to select another tap, then reactivate the power supply. Never rotate the secondary selection switch while the power supply is activated, as this can cause circuit dam-

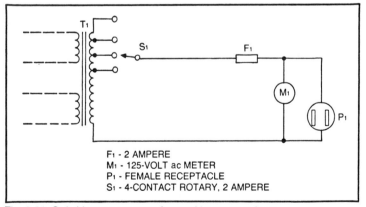

Fig. 6-61. Switching arrangement for varying output voltage.

age. Remember, this power supply should not be activated without some load attached to its output.

PROJECT 26
FREE ELECTRICITY POWER SUPPLY

This next project is quite simple, quite unusual and may be quite useless to a great many readers, but it does make an interesting supply which can be put together for a few dollars and in less than an hour's time. These circuits have been with us for many years and have often been called "electricity stealers", "radio electricity", and a myriad of other names which roughly describe what they do.

While most of us don't think about it, electricity is almost everywhere. It bombards us constantly in the form of commercial broadcast station transmissions. Your radio is a small power supply or power receptor of sorts. Its primary gets its power from the broadcast signals in the air and converts them to audio output. It is actually possible to measure the strength of the broadcast signal with specialized meters. These simply act upon the electricity present and use it to drive miniature meters.

It is possible to use some of this broadcast electricity as the primary driving current for a fairly standard power supply. The circuit is shown in Fig. 6-62 and closely resembles a crystal radio receiver. This is basically what it is, for if you remove the filter capacitors and place an earphone between one of the terminals and ground, you may be able to hear a nearby radio station.

A close examination of the circuit shows it to be a full-wave voltage doubler. You could just as easily have used a half-wave or full-wave bridge circuit, but the voltage doubler will provide a more readable dc voltage at the output.

Normally, the input of the voltage doubler rectifier circuit would be connected to the secondary winding of the transformer. In this instance, however, our transformer secondary is a radio antenna. (I imagine you could consider the radio station's transmitting antenna to be the transformer primary.) Radio frequency energy is a form of alternating current and is picked up on the radio antenna. Here, it is fed to our miniature power supply circuit, which consists of two 50 PIV 1-ampere diodes. (The 1-ampere size was chosen for convenience. Almost any forward current rating will be adequate.) The output from the rectifiers is fed to C1 and C2, which are 500-microfarad, 16-volt electrolytics. Almost any value of capacitance will work here, but both capacitors should be identical. The

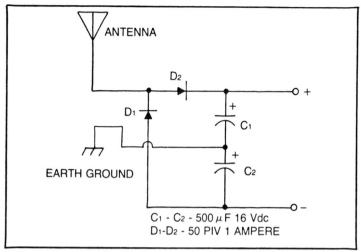

Fig. 6-62. Free electricity power supply.

author chose this value because matching capacitors happened to be available at the time this circuit was built.

Circuit ground is a metal stake driven into the soil (or an alligator clip lead attached to a metal water pipe). This forms the complete circuit. Radio frequency energy which is picked up on the antenna is rectified and doubled by the capacitors and diodes. The resultant output will be about 2.8 times the RMS value of the radio frequency energy.

The power supply components can all be mounted on a small piece of perforated circuit board, paying close attention to component polarity. It is not necessary to mount the circuit board in a compartment of any type, so it may be fitted with screw-in terminals for attachment of the antenna, the ground and for outputting the positive and negative dc values.

The antenna should consist of a length of wire which is at least one-quarter wavelength at the operating frequency of the radio station being received. If this is a nearby AM radio station, best results will be obtained with an antenna which is about 200 feet long. This can be a length of small conductor wire which is strung from a distant tree. Get it as far above the ground as possible. Now, for those readers who cannot accommodate an antenna of this size (and most of us can't), take heart, as almost any reasonable length will work as long as the radio station is nearby. Try to get the end of the antenna as high as possible off the ground and try to obtain as good a ground connection as possible. Don't expect to get any readable results from any but very close local radio stations. The voltage generated at the receiving points by most radio stations is measured in microvolts, which is equivalent to 1/1,000,000th volt. Even with a voltage doubler circuit, we're not talking about a great deal of potential difference.

Again, this circuit will be useless to a great many readers, but for those of you who have a radio or television station within a block or so of your home, you will probably be able to pick up enough rf energy to get some sort of reading at the output of this supply. This circuit was tested within 100 feet of a radio tower and several volts of dc output were measured. This, however, declines significantly as the distance between the antenna and the tower increases.

The only sure way to know that this circuit is working is to attach its output to a dc voltmeter set on the lowest scale possible. If you have a meter which has a millivolt scale, then you are probably in business. Since this device requires no connections to the standard ac line, transport it to a point which is very near a radio tower. Clip a small length of hookup wire to the antenna terminal. Another alligator clip lead will ground the circuit to the automobile body. You should immediately get a reading on the attached voltmeter. If not, move a little closer to the tower.

Better results can be obtained with this simple circuit by adding a tuning arrangement which will allow you to resonate the length of wire used as the antenna. The tuning arrangement shown in Fig. 6-63 will work for all standard AM radio broadcast stations. It consists of L1, which is an AM loopstick antenna. These are availa-

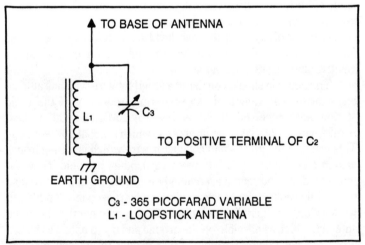

Fig. 6-63. Tuning arrangement for free electricity supply.

ble at most electronic hobby stores or may be salvaged from old pocket receivers. Attach one end of the coil to ground and the other end to the base of the antenna. The antenna base actually starts at the connection between D1 and D2, so this will make an ideal connection point for L1. C3 is a 365-picofarad variable capacitor, which may also be purchased from your local hobby store. This is connected in parallel with L1. A good earth connection at the base of L1 is still desirable. By obtaining an output reading on a meter and then tuning C3, you will be able to determine when the antenna is set to maximum (for a particular radio station frequency) by a peaking of the output dc voltage. Another method of tuning can be attempted using a crystal headphone placed across the positive output terminal and ground. (Disconnect C1 and C2 temporarily while this is being done.) When you hear the desired radio station in the headphone, you know that you're on frequency.

While I suppose it might be possible to actually power some extremely low frequency circuit with this type of power supply, this is highly impractical in most applications. No, this circuit is included here as something that might be interesting and fun for the reader. If you live in a fringe reception area, you may be wasting your time. Those persons who live in close proximity to broadcast stations, however, will get a big kick out of this project when it's completed. Incidentally, if you guess wrong about your ability to pick up free current from the air waves and find that you're just too far away, simply remove D1 and C2 from the circuit along with C1, and use the positive terminal and ground as a connection point for a small earphone. With the loopstick antenna and variable capacitor, you will end up with a crystal radio which requires no electricity to pick up your local broadcast station.

PROJECT 27
28-VOLT POWER SUPPLY USING THREE ZENER DIODES

While the most common low-voltage values required by many types of electronic equipment lie in a range below 18 Vdc, some transistor amplifiers and other devices are designed to operate from 24 or 28 Vdc. Most of these require some form of electronic regulation. The author has found that simple zener diode regulators are often adequate for all but the most demanding circuits.

Unfortunately, while it is easy to obtain zener diodes at values of 6, 9, 12, and 15 volts, it is sometimes difficult (especially in rural areas) to quickly latch onto an inexpensive zener for 28 volts.

Fortunately, this is no real problem, as zener diodes may be combined in series to obtain many different regulating voltages.

Figure 6-64 shows a power supply which uses simple zener diode regulation and produces an output of 27.3 Vdc using common components. The reason the output is not an even 28 Vdc is that three 9.1-volt zener diodes are used in series. They combine to form a single regulator circuit that conducts at any value which is above 3 times 9.1-volts, or 27.3 volts. The 9.1-volt value is fairly common, although three 90-volt units will work just as well.

The supply is very similar to others presented in this chapter, in that it uses a full-wave bridge rectifier assembly acting upon the entire transformer secondary value. The rectified current is filtered by C1, which is a 500-microfarad unit. The regulator circuit is composed of the series resistor, R1, and the three series-connected zener diodes. The schematic shows that chassis connections are indicated for ground. However, this supply may be isolated above chassis ground by simply interconnecting the back of the bridge rectifier, C1's negative terminal, and the bottom of CR3 together. This lead will then serve as the negative output.

Figure 6-65 shows a suggested component parts layout on a medium-sized section of perforated circuit board. As with some previous projects, the transformer and primary circuit are installed in a small plastic or aluminum container. The circuit board is wired and then mounted in close proximity to the transformer secondary. Be sure of your connections in the zener diode string, as a reversal here will make the regulator circuit useless. You will still read an output voltage, but it will not be regulated.

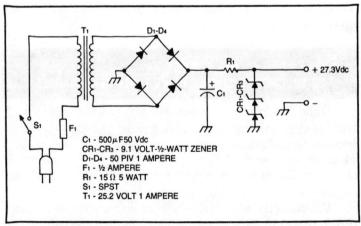

Fig. 6-64. 28-volt power supply using three series-connected zener diodes.

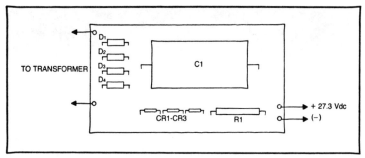

Fig. 6-65. Parts layout for 28-volt power supply.

Once the circuit is complete and your visual inspection indicates that there are no wiring errors, connect a dc voltmeter across the output. Make certain that it is adjusted to a range which will read the 27.3-volt output (0-50 Vdc). Activate the supply by engaging S1 and note the output reading. You may get a slightly different reading due to inaccuracies within your meter, but the output should be within a volt or two of nominal. If you get no output reading, start at the primary circuit and trace through to the output for the problem or problems. On the other hand, if the output reads closer to 35 volts, this is an indication that your regulator circuit is inoperational. This will be most likely due to a reversed or defective zener diode in the string. Check each diode and replace any that are questionable.

There is nothing highly unusual about this straightforward power supply other than the fact that we have used three zener diodes in series to arrive at the desired regulating value. We might just as easily have used a single zener diode rated at 28 Vdc, but this project assumed that the 9-volt unit could be more easily (and inexpensively) obtained.

If you have a circuit or device which requires 24-volt operation, this same circuit may be used but is modified by replacing CR1-CR3 with two 12-volt zener diodes wired in series. This will yield a single diode which will conduct at any value above 2 times 12 volts, or 24 Vdc. You can combine as many diodes in series as you want to arrive at higher and higher regulating values. Don't try a parallel connection, however. Some persons may think that two 6-volt zener diodes wired in parallel will yield the equivalent of a 3-volt zener. This is not true. In the above instance, the circuit will still regulate at 6 Vdc. If a 3-volt and 12-volt unit were wired in parallel, then the low voltage zener would control the circuit and would conduct when the output value exceeded 3 Vdc.

PROJECT 28
USING A CURRENT METER TO MEASURE VOLTAGE

Many times it is desirable, if not mandatory, to meter the output voltage of a dc power supply. This is a relatively simple matter and involves placing the positive and negative meter contacts across the output of the power supply. All of the projects in this book have been checked out with the aid of an external dc voltmeter. This will usually be a multimeter, one which will measure voltage (ac and dc), current, resistance, etc.

Sooner or later, you will want to incorporate a built in meter as part of the power supply circuitry. This will mean that a mounting hole will be cut in the power supply chassis or cabinet, the meter installed and its two terminals attached to the power supply circuitry. This will usually require the purchase of a voltmeter from a hobby store or electronics outlet. Naturally, the meter must have a range which is adequate and *proper* to measure the output potential of the power supply.

Metering is sometimes the most critical portion of power supply circuitry. This is especially true when output voltage must be kept to within very close specifications. Also, certain high voltage power supplies must be metered when used to provide operating potential to radio frequency amplifiers. These devices usually must be held to within a certain power output. This is determined by multiplying the plate voltage times the plate current. Obviously, if your metering is defective and giving erroneous readings, the amplifier could possibly be operating at illegal outputs without the operator ever being aware of the situation.

When we approach high voltage potentials, meters become more and more expensive. It is not unusual for a commercial meter designed to read a maximum of 4000 Vdc to cost in excess of $150. The actual price will depend upon meter construction and overall accuracy.

Suppose you have a power supply with an output of 150 Vdc. One might assume that any meter with a scale which extends to 150 or more volts is adequate to measure this potential. This is basically a true statement as far as *adequacy* is concerned. It is not enough, however, to possess a meter with an adequate scale. The scale must also be of the *proper* range for accurate measurements.

To explain the difference between accurate and proper, let's take our example of a 150 Vdc power supply whose output voltage we desire to measure. Assume that a 150 Vdc meter is used (one which will read a maximum of 150 Vdc). Now, when the power supply is activated, the 150-volt output will drive the meter indicator to full scale. If the power supply output drops to 145 volts, we will see the needle drop back a bit. But what if the power supply output voltage climbs to 155 Vdc? The meter will not indicate this, since its top scale reading is only 150 volts. It can be seen that the meter scale must be rated for a value higher than the maximum output potential of the power supply circuit to which it is connected.

Let's take another example using the same 150 Vdc power supply. Suppose you use a meter designed to measure a maximum potential of 1500 Vdc. This means that a 1500-volt output will drive the meter to full scale deflection. But when a potential of 150 Vdc is connected to this meter, the needle indicator will only climb about 10 percent of the way up the scale. All readings will be confined to the first 10 percent of the meter face and the remaining 90 percent will be unused. This presents several problems. First of all, it is quite difficult to take readings from a meter which is deflected so little. Think of it. If a 10 percent deflection is had with an input of 150 volts, a decrease of 15 volts will cause the meter indicator to drop a distance equal to 1/100th of the entire scale. The result is that it is nearly impossible to see the true reading. Secondly, from an accuracy standpoint, most meters provide the truest readings when their indicators are deflected to the upper 50 percent of scale. In other words, the 1500 Vdc meter, to be most accurate, would need to be driven by an input voltage of over 750 Vdc in order to be deflected to the top half of the scale.

Both examples of meters for connection to our 150 volt circuit may be adequate, but they are certainly not proper for the potentials dealt with here. A good rule of thumb when choosing meters is to opt for one that, when driven by your power supply, will have a meter deflection at the three-quarter point on the scale. For a 75-volt power supply, a 100-volt meter would be ideal, and for the 150-volt power supply discussed as an example here, a 200-volt meter would be best. This assumes that 150 Vdc is a nominal voltage output and takes into account that the voltage could increase by a maximum of 15 percent and decrease by approximately the same amount.

This brings us back to the situation at hand and the problems associated with obtaining the correct meter after the correct scale is known. While it is true that you can get a meter equipped with about any scale desired, these are sometimes terribly expensive, especially when they are designed to measure other than standard values around which most meters are designed. Looking through electronics supply catalogs, you will commonly see voltmeters designed with maximum scales of 3, 5, 10, 15, 25, and 150 Vdc. You may run into a few other values, but these are the most common. Some of these may cost less than $10, but most will be in excess of $15 or $20. Obviously, if you've gone to the trouble of building your own power supply from scratch in order to be economical, you probably won't want to spend $20 on a component which may cost more than the entire power supply proper. This is often the case in power supply construction because most of the components cost very little, especially when they can be salvaged from other equipment.

Since you built the power supply yourself, why not do the same thing with the power supply meter? No, it's not practical to actually build a voltmeter from scratch, but if you look hard enough, you can certainly find a surplus meter for a few dollars, or you may even be able to pull one out of some old equipment. Chances are, however, the scale on the meter will not be adequate or proper for the power supply potential you need to measure. Here's where the building comes in.

I have found that war surplus catalogs quite frequently contain very inexpensive meters originally used to measure current. Actually, all a voltmeter is is a microammeter or milliammeter with a series resistance. So, if you can obtain a meter designed to measure milliamperes or microamperes, chances are you can convert it to a functioning voltmeter for a couple of dollars.

Figure 6-66 shows a sample circuit based around a 0-50 microampere meter, which is a quite common value. Without going into complicated mathematics, the addition of a 1-megohm resistor will change this to a voltmeter designed to read 0-50 Vdc. Each time the series resistor is increased by a factor of ten times, the voltage scale is multiplied by ten. As is shown in the schematic drawing and components list, a 10-megohm resistor will produce a meter with a 500-volt scale, and 100 megohms takes us all the way up to 5000 Vdc. While this circuit uses a 0-50 microammeter, a 0-100 microammeter would still be affected in the same manner. A 1-megohm resistor in this case would give you a 100-volt scale which would be increased by ten times each time the series resistor is increased by the same factor.

War surplus catalogs often contain 0-500 microammeters for a few dollars. To figure the value of the series resistance here, simply divide the resistances in Fig. 6-66 by ten and you'll come up with the

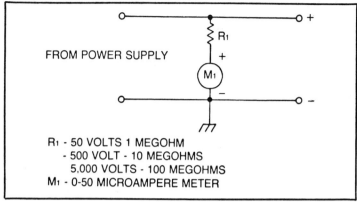

Fig. 6-66. Simple metering circuit using a 0-50 microammeter.

equivalent values for this meter designed to read a maximum of 500 microamperes. It can be seen that a 100-kohm resistor will produce a voltmeter with a 0-50 Vdc scale. A 1-megohm resistor will yield a 500 Vdc scale, etc.

For truly accurate readings, it is best to use high grade resistors. These usually have a tolerance of 1 percent or less. Using these resistors, the reading obtained on your meter could be as much as 1 percent off because the resistor could differ from this value by that much. Standard grade resistors often have 5 percent and 10 percent tolerances and are not as desirable.

When dealing with precision resistors, it is often impossible to purchase single units which provide enough ohmic value. Precision resistors may be available with values up to 1 megohm, but most stop around 500 kohms. Obviously, it will be necessary to wire series strings of these resistors to arrive at the desired value. Remember, resistors in series add, so ten 500,000-ohm resistors (500 kohm) will yield a total resistance of 5,000,000 ohms (5 megohms).

Caution: When designing a meter to read high voltage, never use a single series resistor, even if one of the proper value is available. Most ½-watt resistors are incapable of handling high voltage potentials. Here, it is best to use eight or ten separate components in series. This avoids flashover, which almost always results in permanent meter damage. The series resistor string can be wired on a small section of perforated circuit board and attached directly to the back of the meter using its terminals for insulation and support. Remember, the full power supply voltage potential is seen at several points on this circuit board and must be properly insulated and shielded at high voltage potentials to prevent a possible shock hazard.

For extremely accurate measurements, it may be necessary to align this metering circuit by means of an external meter of known accuracy. The circuit in Fig. 6-67 will allow for precise meter adjustments and incorporates a precision variable resistor in series with R1. The value of this variable resistor will depend upon the type of meter used and the final scale you need to end up with, owing to the power supply voltage. Don't use a common potentiometer, as this will provide poor adjustment possibilities and even poorer accuracy. Rather, a precision, multi-turn, wire-wound trimmer resistor should be installed, one that can be adjusted by means of a miniature screwdriver. These usually cost less than three dollars, even at commercial prices. Read the output from your power supply on an external meter of known accuracy, then adjust the variable resistor until the internal meter is aligned with the external one. It may be necessary to re-check the alignment every year or more often if the power supply is in continuous use and output voltage is critical.

Nearly any type of microammeter or milliammeter can be called to service as an accurate voltmeter when fitted with the proper series resistors. The author prefers the microammeters because they are a little more versatile and may also be used in a switching network to measure current. But the milliammeters may often be cheaper and are available through more surplus outlets.

Incidentally, voltmeters which may be on hand can also be modified to measure higher voltage values by using series resistors. For example, a 0-10 Vdc meter can be modified to read 0-100 Vdc, but it will be necessary to arrive at the resistor value by trial and error. The actual value will depend upon the internal meter resistance, which can vary from unit to unit. Start here with a very high resistor value and gradually lower it until the meter indicator comes in range. A high value variable resistor is helpful here and need not be of the precision value, since it will not remain in the circuit upon completion. Insert the resistor between the meter and the power supply output and adjust it until you get a fairly accurate indication. Then, remove the variable resistor without changing its setting and measure its internal resistance with an accurate ohmmeter. Once you know the value, a precision fixed resistor of the same value may be substituted. It will still be necessary, in most instances, to include a precision trimmer resistor to obtain the best possible accuracy.

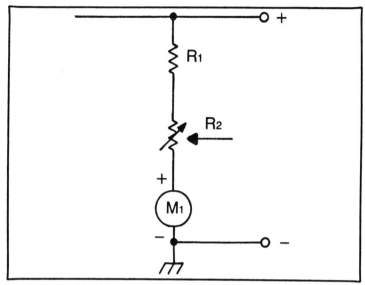

Fig. 6-67. Circuit which allows precise meter adjustment.

All in all, this type of meter modification is very simple and inexpensive, owing to the minimum of components needed and the possible use of salvaged parts. Such a circuit can be completed in less than an hour and will be far less expensive than purchasing a commercial meter. It may be more accurate too. This will depend upon the quality of surplus meter you are able to obtain and on the tolerance of the series resistors.

PROJECT 29
A DIFFERENT TYPE OF VARIABLE SUPPLY

Most of the power supplies discussed in this chapter and which produce variable outputs (nonregulated) have accomplished this by placing a variable resistor in series with the secondary output. There is nothing particularly wrong with this type of control, but it is not the only way to provide variable voltage output nor even the most desirable in some applications. Sometimes, it is preferable to vary the transformer input voltage, especially when secondary current may be very high while the voltage is low. For example, a 6-volt power supply which delivers 5 amperes to the load draws only 30 watts of power breaks down to a current consumption in the transformer primary of approximately 250 milliamperes. It is much easier to control current in this range with a series resistance than the higher amount.

Figure 6-68 shows a supply primary circuit which has been fitted with a variable resistor. No component values are given here, because they will vary with the amount of current drawn in the primary circuit. Given the example of a 6-volt supply delivering 5 amperes and a current consumption of 250 milliamperes in the primary, the primary voltage will drop 25 Vac for every 100 ohms of resistance. This means that a resistance value of 200 ohms will reduce primary voltage by nearly half. The secondary voltage will do likewise. Fortunately, it will probably not be necessary to bring about this amount of voltage drop at the secondary, and a 0-100 ohm variable resistor will be all that is required. This will allow for a 25 percent decrease in primary voltage at the maximum setting. This will drop the secondary voltage to 4.5 Vdc. Using this type of circuit, the secondary voltage can be quite accurately set without inserting any series resistance in the high current secondary output.

The value of the primary series resistor is determined by Ohm's Law: $E = IR$, where E is the voltage drop, I is the known primary current in amperes, and R is the series resistance. Since we must figure for R, $R = \frac{E}{I}$. This means that if we want a maximum voltage drop of 25 Vdc in the primary and know the primary current to be 250 milliamperes, the formula will read: $R = 25/.250$ or 100 ohms. The power rating of the series resistance is figured by the Ohm's Law power formula of $P = I^2R$. Figuring a maximum resis-

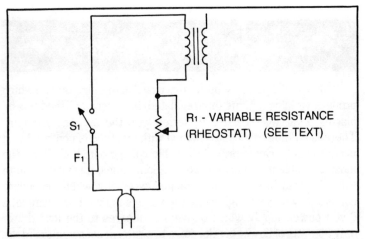

Fig. 6-68. Variable voltage circuit using variable resistor in the primary line.

tance of 100 ohms, we arrive at an answer of 6.25 watts. Since no variable resistors are made for this value, a 10-watt component would most likely be chosen.

There are no complex modifications required to include variable primary control. Simply drill a hole in the power supply chassis to accept the variable resistor, break the primary circuit for its insertion and you're finished. The cost of this project will be the same as that of the variable resistor, which will most likely be a rheostat that can be purchased from most commercial mail order catalogs. A 10-watt unit should cost less than eight dollars. These devices are commonly available with ratings up to 300 watts. Even the high powered type will cost less than thirty dollars.

PROJECT 30
MODIFYING A LOW VALUE
CURRENT METER TO READ HIGH VALUES

Often, it is difficult to obtain an ammeter (milliammeter or microammeter) which is of suitable range to take care of power supply output metering. This is a similar problem to the one discussed in the previous project; but here, the insertion of a series resistor is not practical. It is quite practical, however, to modify any meter designed to read current of one value to enable it to read current of a much higher value. For example, a 0-50 microammeter can be modified to read 0-500 microamperes.

A sample circuit is shown in Fig. 6-69. Notice that the meter is placed in series with the positive output leg. However, a parallel resistance has been added, Rs. This is a shunt resistance which is effectively in parallel with the internal resistance of the meter. It is very difficult to determine the internal resistance because it is of a very low value in most instances. If you place the probes of an ohmmeter across the meter terminals in an attempt to measure internal resistance, high current will flow within the circuit and the meter may be damaged permanently.

I usually assume the resistance to be less than 1 ohm. This will be true in most instances but is not a hard and fast rule. Thus, it is necessary for the shunt resistance to be equal to this value in order to effectively double the range of the meter. As the shunt resistance decreases in value, more and more current flows to this portion of the circuit rather than through the meter. If you have a 0-50 microammeter, the trick is to make the shunt resistance small enough to pass all of the output current except 50 microamperes maximum.

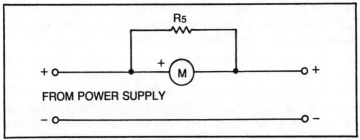

Fig. 6-69. Metering circuit with external shunt resistor.

If the output of the power supply is approximately 500 microamperes, then the shunt must be adjusted so that it will pass 450 microamperes when the total power supply output is 500 microamperes. This allows 50 microamperes to flow through the meter, so a reading of "50" on its face will now indicate 500 milliamperes.

The shunt resistor is not a standard type. It is one you make yourself. The easiest way to accomplish this is to purchase a packet of surplus, single-layer rf chokes. You will find that their resistance will usually be less than 1 ohm (often, a small fraction of an ohm) and will, of course, vary depending upon the length of conductor used to wind the device. Next, you must build a source of known current. This can be done quite effectively using a 1.5-volt D-type battery and a resistor. By wiring the resistor in series with the battery output and using Ohm's Law, you can determine the output current by the formula $I = \frac{E}{R}$. This is where I is current measured in amperes, E is the battery voltage (1.5 Vdc), and R is the resistance. For example, if the resistor is rated at 10 ohms, the current read by an external milliammeter will be .15 ampere, or 150 milliamperes. If you would care to use two flashlight batteries in series, the output would be twice this amount, or you could replace the resistor with a 5-ohm value to achieve the same results.

Temporarily connect the choke (or anything else you use as an extremely low value resistor) in parallel with the meter. Gently brush the output leads of the current source across the meter terminals. (Be sure to observe polarity.) Do not provide a firm connection at this time! If your meter does not respond at all, then the resistance is too low and is passing nearly all of the current. If, however, the meter needle pegs itself, then your resistance is too high and some turns must be removed from our makeshift wirewound resistor to decrease it so that less current will flow through the meter. Keep removing turns until the meter indicates the known current value. Very precise adjustments can be made by slowly clipping small sections from the wire used to make the resistor.

If you don't wish to use surplus chokes, buy a reel of the smallest enamel-clad wire you can find. The smaller the wire, the less length required to build up a considerable resistance. The body of a ½ watt carbon resistor, along with its leads, may be used as a perfect mounting platform. Choose a high resistance value if you choose to take this route. All in all, however, you will probably find that using surplus single-layer chokes which can be had for a few cents each will be the most economical method of obtaining the correct shunt resistance.

Using this modification principle, almost any current meter can be used to measure values which would far exceed its range under normal circumstances. Warning: When wiring the shunt for very high current circuits, it is necessary to use conductor which is rated to handle the current. Here, it may be necessary to use larger diameter conductor and a lot more of it to establish the proper resistance.

PROJECT 31
VOLTAGE AND CURRENT READINGS WITH A SINGLE METER

Since even surplus meters can become rather expensive, if we could use one meter to measure both voltage and current, some cost savings would certainly be enjoyed. Additionally, many low voltage power supplies are extremely small in physical size and will not easily accept the installation of two separate meters. This is not too difficult to accomplish, since you will notice that your multimeter already does this. You can take voltage, current, and resistance measurements all with the same meter. The circuit shown in Fig. 6-70 will allow you to read the voltage and current of a power supply by simply throwing a switch.

As shown, the circuit should read up to 50 Vdc and can measure current as high as 5 amperes, although the latter is not recommended due to the small size of the shunt. This will work best with power supplies designed to deliver 2 amperes of current or less. If higher current applications are required, it would be best to wind the shunt with larger wire but still arrive at the same resistance value.

The output from an unmetered power supply is fed to this simple circuit, which can also be made a part of the supply proper.

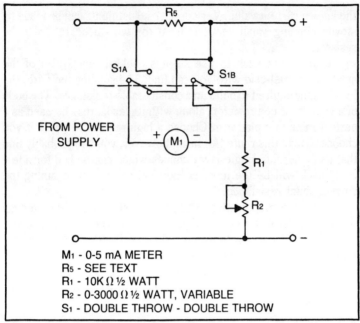

M₁ - 0-5 mA METER
R₅ - SEE TEXT
R₁ - 10K Ω ½ WATT
R₂ - 0-3000 Ω ½ WATT, VARIABLE
S₁ - DOUBLE THROW - DOUBLE THROW

Fig. 6-70. Metering circuit allows for voltage and current readings with a single meter.

Just before the output terminal of the unmetered supply, break the positive circuit lead and insert Rs, which is a shunt resistance. The author made his from approximately 2.5 inches of #28 enamel-clad wire. Some experimentation will have to be made in order to arrive at a meter which will read 0 to 5 amperes. Test this out using the methods described in the previous project.

Switch S1 places the shunt in the circuit when thrown to the top position and removes M1 from the voltage measuring mode, which is selected when S1 is in the down position. The positive and negative terminals of the meter are connected to the switch arm as shown. The voltage measuring circuit consists of R1 and R2 in series with the milliammeter. R2 should be a precision unit which can follow for precise alignment of this meter with a known current source.

The only complicated hookup is in making certain that the right contact points are brought into S1. This is a double-pole double-throw type and acts like two SPST switches operated in parallel and simultaneously.

Make certain that the polarity of M1 is closely observed and no problems should result. The entire circuit may be constructed in an aluminum box separate from the power supply but will most often be used as part of one of the low voltage power supply projects already discussed. Using this method, both voltage and current can be read by using a single meter. You are certainly not restricted to using a 0-5 milliampere meter as specified in this discussion. Almost any type will work, but it will be necessary to alter the values of the shunt resistor, as well as R1 and R2, as described in the previous two projects.

PROJECT 32
ALTERING SECONDARY VOLTAGE

Sometimes, when salvaged transformers are used for dc power supplies, you will find that the output voltage of the secondary may be a bit high or low. When the deviation in output voltage measured at the secondary is only slightly high or low compared with the ideal value, sometimes an easy adjustment can be had. This is especially true of power transformers which have a medium or high voltage winding and one or more low voltage windings.

Several previous projects discussed combining transformer primaries and secondaries in series-aiding to increase output voltage. These projects used two or more transformers. Figure 6-71 shows how a single transformer with a complex secondary winding can be wired to increase or decrease the secondary output voltage.

Let's assume that the medium voltage winding reads 520 volts. Assume further that the desired secondary output should be of at least 530 volts. If your transformer has a 12-volt filament winding, then by wiring it in series-aiding as shown in Fig. 6-71A, the total secondary output will be 520 Vac plus 12 Vac, or 532 volts.

Suppose the 520-volt secondary output is a shade too high. By connecting the windings in the series-opposing mode as shown in Figure 6-71B, the low voltage will be subtracted from the high one, and the total output will be 520 Vac minus 12 Vac, or 508 volts.

Some transformers may have two or more low voltage windings. This is especially true of old television transformers which usually have a medium voltage winding, a 12 or 6-volt winding, and a 5-volt winding (originally designed for the rectifier tube filaments). Some may even have two 12-volt windings and one 5-volt winding. Obviously, those transformers with a large number of secondary windings will be capable of more output voltage modification.

Although it is rare, occasionally a transformer will be seen which contains a low voltage secondary designed to deliver less

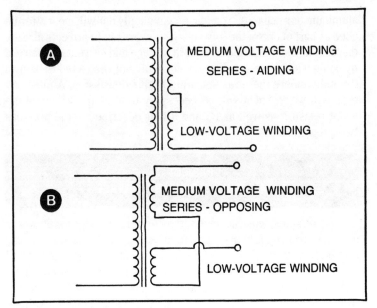

Fig. 6-71. Technique for altering secondary voltage.

current than the medium or high voltage windings. While most filament windings deliver an ampere or more of current, this is not always the case. In any event, the series windings that are added to the medium voltage winding must be capable of withstanding a current flow equal to that of the medium voltage winding. Incidentally, if your transformer only has a single medium voltage secondary output, you can always wire the secondary of a separate 6 or 12-volt transformer in series-aiding or series-opposing for the same results. Chances are, the determination of whether the windings are connecting in the aiding or opposing configuration must be made by using an external voltmeter to measure voltage output after the connections have been made. If you are seeking series-aiding and end up with a decrease in voltage, simply reverse the low voltage winding connections, and the opposite configuration will be brought about.

A means of secondary voltage control can be had by combining transformer windings in series-aiding and series-opposing. However, the amount of control is usually limited to less than 20 volts, either aiding or opposing. This is due to the value of the low voltage secondary windings. If there were some way whereby this low voltage could be used to bring about a larger change, a single transformer with a multiple secondary winding could be made even more versatile. Since the secondary output voltage in a medium voltage transformer is a multiple of the primary input voltage value, when the primary voltage is changed, the change is multiplied (either up or down) at the secondary. For instance, if the secondary medium voltage output is five times that of the primary input voltage, then a reduction of 10 volts at the primary will result in a reduction of 50 volts at the secondary.

Figure 6-72 shows how a low voltage secondary winding may be wired in series aiding with the primary winding in order to bring about an effective decrease in secondary output voltage at the medium voltage winding. Assuming that the primary is designed to operate from 115 Vac and that low voltage secondary winding is 12 Vac, when the two are combined in series-aiding, they form a primary winding (complex) which is designed to operate from 115 Vac plus 12 Vac, or 127 volts. If the primary supply is still delivered at 115 Vac, then the complex winding is receiving only 90 percent of the required voltage to produce a full output. If the medium voltage winding normally produces an output of 500 volts, this output will now be 450 volts (90 percent of 500).

Again, the more secondary low voltage windings contained on the transformer, the more versatile the voltage control possibilities become. And as with the previous project, separate low voltage transformers may be used to bring about the voltage change. Series-opposing may also be used in this method to increase the medium voltage output. This is the opposite of the previous project, where series-aiding increased output voltage and series-opposing reduced it. In this circuit, it can be said that series-aiding increases the needed input voltage to produce a certain output, while series-

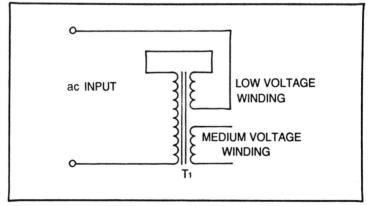

Fig. 6-72. Decreasing secondary voltage through a series-aiding arrangement.

opposing decreases the needed value of input voltage. Since input voltage remains the same in either case, a needed increase results in a lower secondary output, while a needed decrease at the primary results in an increase at the secondary.

SUMMARY

The 32 projects in this chapter have been chosen to provide you with a well-rounded sampling of circuits which can be put to many useful applications. Some projects may be assembled in less than an hour and for a few dollars, while others may cost $100 or more. All represent a very distinct savings when compared with commercially manufactured power supplies that are sold on today's electronics market.

Many of the projects may have immediate application for you in your experimental endeavors, while others may be tucked in the back of your mind to be assembled at some later date for a specialized purpose. All have not one use but many.

As a final note, I urge the strict adherence to safety procedures when building any electronics project. Even a slight burn from a soldering iron can handicap your building progress for many days. Be especially careful when building and testing the medium and high voltage power supplies in this book, as a second's forgetfulness here can cost you your life. I had two good friends who were engineers at commercial radio stations, each having more than twenty years electronics experience. Both were killed (on separate occasions) by coming into contact with high voltage power supplies which were supposed to be turned off and completely discharged. Neither was. There is nothing inherently dangerous about electronics building and experimentation. Like many other pursuits, however, the electronics field is sometimes very unforgiving of mistakes.

A good basis has been laid throughout this book in power supply technology and design. The reader is encouraged to substitute applicable parts where practical and to experiment as much as possible. As long as your power supply is properly fused at the primary circuit, there is little chance that you will do any permanent damage to most power supply components due to substitution of an improper part. This, of course, does not apply to complex regulated supplies, which may use expensive series transistors. As you gain more and more expertise in building, testing, and operating dc power supplies, the selection of substitute parts will come quite easily and you will be able to custom design those circuits which are best suited to your interests.

Appendix A
Schematic Symbols

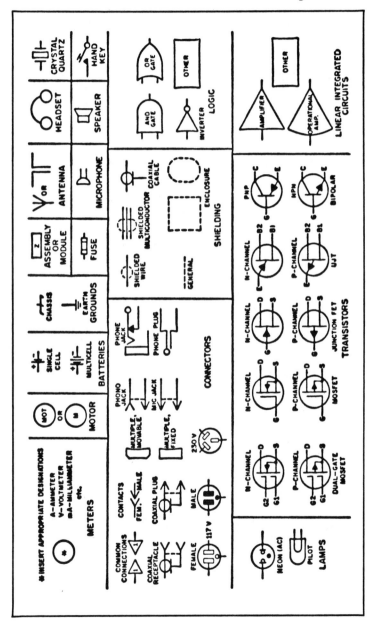

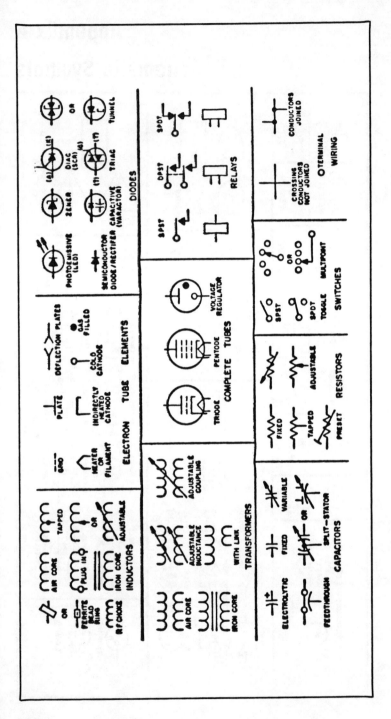

Appendix B
Wire Size and Current Rating

Length (ft) →	30	40	50	60	70	80	90	100	125	150	175	200	225	250
Ampere Load								Wire Sizes						
5	14	14	14	14	14	14	14	12	12	10	10	10	8	8
6	14	14	14	14	14	12	12	12	10	10	10	8	8	8
7	14	14	14	14	12	12	12	12	10	10	8	8	8	8
8	13	13	13	12	12	12	12	10	10	8	8	8	8	6
9	14	14	12	12	12	12	10	10	8	8	8	8	6	6
10	14	12	12	12	10	10	10	10	8	8	8	6	6	6
12	14	12	12	10	10	10	10	8	8	6	6	6	6	4
14	14	12	12	10	10	10	10	8	8	6	6	6	6	4
16	14	12	10	10	10	10	8	8	6	6	4	4	4	4
18	12	12	10	10	10	8	8	8	6	6	4	4	4	4
20	12	10	10	10	8	8	8	6	6	4	4	4	4	4
25	12	10	10	8	8	8	6	6	6	4	4	4	4	2
30	10	8	8	8	6	6	6	4	4	4	2	2	2	2
35	10	8	8	6	6	6	4	4	4	2	2	2	2	1
40	8	8	6	6	6	4	4	4	2	2	2	2	2	2
45	8	6	6	6	4	4	4	4	2	2	1	1	1	1
50	8	6	6	4	4	4	4	4	2	2	1	1	1	1
60	6	6	4	4	4	2	2	2	2	2	1	1	1	1
70	6	4	4	4	2	2	2	2	1	1	0	0	0	0
80	6	4	4	3	3	3	2	1	0	0	0	0	0	0
90	4	4	2	2	2	1	1	0	0	0	0	0	0	0
100	4	2	1	1	1	0	00	00	000	000	000	000	000	00
130	4	2	0	0	00	00	00	000	000	0000	0000	0000	0000	000
160	4	2	1	0	00	000	000	0000	0000	250	300	250	250	250
200	2	1	0	0	00	000	000	0000	250	300	350	300	300	300

Appendix C
Resistor Color Code

Color	Significant Figure	Decimal Multiplier
Black	0	1
Brown	1	10
Red	2	100
Orange	3	1,000
Yellow	4	10,000
Green	5	100,000
Blue	6	1,000,000
Violet	7	10,000,000
Gray	8	100,000,000
White	9	1,000,000,000
Gold	—	0.1
Silver	—	0.01

Index

A

Ac, 2
Ampere, 3
Average forward current, 17
Average forward voltage drop, 10
Average rectifier forward current, 11
Average reverse current, 11

B

Battery charger, power supply, 170
Bleeder resistor, 88
Bleeder resistor voltage divider, 26
Bridge rectifier, 16

C

Capacitance, 60
Capacitor, ceramic, 74
Capacitor, charging a, 63
Capacitor, discharging a, 67
Capacitor, electrolytic, 74
Capacitor, mica, 73
Capacitor, oil, 74
Capacitor, paper, 72
Capacitor color codes, 76
Capacitor input filter, 22
Capacitor insulation, 86
Capacitor plate spacing, 69
Capacitors, for power supplies, 77
Capacitors, types of, 72
   fixed, 72
   rotor-stator, 72
   trimmer, 72
   variable, 72
Capacitors in parallel, 79
Capacitors in series, 81
CCS, 45
CEMF, 66
Choke input filter, 23
Choke input filter, multiple section, 24
Color code, resistor, 288
Combining transformers, 48
Counter electromotive force, 66
Cross-referencing, 119
Current, 3
Current meter, 277
Current regulator, 115

D

Dc, 2
Dc, pulsating, 5
Dc blocking voltage, 10
Dc power supply, 1
Dc to dc power supply, 252
Dielectric, 62
Dielectric constant, 71
Dielectric material, 70
Diode rectifier circuits, 11
Diodes, 9
Diodes, rectifier, 10
Diode specifications, 10
Displacement current, 65
Dry cell replacement power supply, 196

Dual-polarity regulated 15-volt supply, 158
Dual voltage power supply, 153

E

Electromotive force, 65
Electron tube shunt voltage regulator, 108
Electronic components, 127
EMF, 65
Experimenter's "junk box", 120

F

Filter, capacitor input, 22
Filter, choke input, 23
Filter, pi, 24
Filter, rc capacitor input, 26
Filters, power supply, 20
5-Vdc, 1-ampere IC power supply, 199
Free electricity power supply, 264
Full-wave bridge, 28
Full-wave bridge supply, 146
Full-wave center-tapped supply, 142
Full-wave combinational, 30
Full-wave rectifier, 14
Full-wave voltage doubler, 33
Full-wave voltage tripler supply, 180

H

Half-wave power supply, 136
Half-wave rectifier, 12
Half-wave voltage doubler, 32
Hertz, 2
High voltage dc power supply, 232
High voltage power supply, 244

I

ICAS, 45
IC controlled variable voltage supply, 211
Integrated circuit, 98
Inverter, 12-volt, 257

J

Junction diodes, 17

K

Kilovolt, 2

L

Light-sensitive solid-state devices, 100

M

Maximum power dissipation, 11
Megavolt, 2
Meter, current, 277
Meters, 270, 279
Microampere, 3
Milliampere, 3
Millivolt, 2
Multi-output add-on regulator, 187

N

9-volt series-regulated supply, 163
Nominal zener breakdown, 11

P

PCB, 126
Peak reverse voltage rating, 54
Peak surge current, 11
Pi filter, 24
Plates, 62
Power supply, dc, 1
Power supply, dc to dc, 252
Power supply, dry cell replacement, 196
Power supply, dual-polarity, 158
Power supply, dual voltage, 153
Power supply, 5-Vdc, 1-ampere IC, 199
Power supply, free electricity, 264
Power supply, full-wave bridge, 146
Power supply, full-wave center-tapped, 142
Power supply, full-wave voltage tripler, 180
Power supply, half-wave, 136
Power supply, high voltage, 244
Power supply, high voltage dc, 232
Power supply, IC controlled variable voltage, 211
Power supply, 9-volt series regulated, 163
Power supply, series-regulated dual-polarity, 202
Power supply, 600 volt, 209
Power supply, solar, 216
Power supply, surge protection of, 228
Power supply, switchable full/half-voltage, 183
Power supply, 300 volt, 206
Power supply, transceiver, 190
Power supply, 28-volt, 267
Power supply, variable, 275
Power supply, voltage doubler, 174
Power supply battery charger, 170
Power supply components, 39
Power supply construction, terminal strip method, 139
Power supply filters, 20

Power transformers, 18
Protective components, 58
Pulsating dc, 5
Pulsating dc waveforms, 7

R

Rectification, 5, 8
Rectifier, bridge, 16
Rectifier, full-wave, 14
Rectifier, half-wave, 12
Rectifier diodes, 10
Rectifiers, solid-state, 54
Regulator, current, 115
Regulator, multi-output add-on, 187
Regulator, series, 109
Regulator, zener diode, 149
Repetitive forward current, 17
Repetitive reverse voltage, 17
Resistor, bleeder, 88
Resistor color codes, 288
Root-mean-square, 4

S

Schematic symbols, 285
SCR, 93
Secondary voltage, altering, 281
Semiconductor devices, 90
Series-regulated dual-poalrity supply, 202
Series voltage regulator, electron tube, 110
Shunt regulator, 103
Silicon-controlled rectifier, 93
Sinusoidal, 4
600 volt power supply, 209
Solar cells, 101
Solar power supply, 216
Solid-state devices, light sensitive, 100
Solid-state rectifiers, 54
Solid-state shunt voltage regulator, 107
Surge current, 17
Surge protection for power supplies, 228
Switchable full/half-voltage power supply, 183

T

Tapped primary windings, 46
300 volt power supply, 206
Transceiver power supply, 190
Transformer power ratings, 42
Transformers, 39
Transformers, combining, 48
Transformers, power, 18
Transformers, series and parallel connection of, 50
Transistors, 91
Triac, 96
12-volt inverter circuit, 257
28-volt power supply, 267

V

Variable power supply, 275
Versatile voltage doubler supply, 174
Volt, 2
Voltage, 2
Voltage, altering secondary, 281
Voltage divider, bleeder resistor, 26
Voltage multipliers, 31
Voltage quadrupler, 37
Voltage regulator, electron tube shunt-detected series, 114
Voltage regulator, shunt, 107, 108
Voltage regulator, shunt detected series, 111
Voltage regulator, solid-state series, 109
Voltage regulator, solid-state shunt-detected series, 111
Voltage regulators, 103
Voltage tripler, 34

W

Waveforms, pulsating dc, 7
Windings, tapped primary, 46
Wire current ratings, 287
Wire size, 287

Z

Zener diode, 11
Zener diode regulator, 149
Zener diodes, 11
Zener voltage regulator, 104